An Introductory Course in Commutative Algebra

An Introductory Course in Commutative Algebra

A. W. Chatters

School of Mathematics
University of Bristol

and

C. R. Hajarnavis

Mathematics Institute
University of Warwick

Oxford • New York • Tokyo
OXFORD UNIVERSITY PRESS
1998

Oxford University Press, Great Clarendon Street, Oxford OX2 6DP

Oxford   New York

Athens   Auckland   Bangkok   Bogota   Bombay
Buenos Aires   Calcutta   Cape Town   Dar es Salaam
Delhi   Florence   Hong Kong   Istanbul   Karachi
Kuala Lumpur   Madras   Madrid   Melbourne
Mexico City   Nairobi   Paris   Singapore
Taipei   Tokyo   Toronto   Warsaw

and associated companies in
Berlin   Ibadan

Oxford is a trade mark of Oxford University Press

Published in the United States
by Oxford University Press Inc., New York

A catalogue record for this book is available from the British Library

Library of Congress Cataloging in Publication Data
Chatters, A. W.
An introductory course in commutative algebra / A. W. Chatters and
C. R. Hajarnavis.
Includes bibliographical references and index.
1. Commutative algebra.   I. Hajarnavis, C. R.   II. Title.
QA251.3.C53 1998   512'.24–dc21   97-49207
ISBN 0 19 853423 X (Hbk)
ISBN 0 19 850144 7 (Pbk)

Typeset by Technical Typesetting Ireland, Belfast, Northern Ireland
Printed in Great Britain by Biddles Ltd., Guildford & King's Lynn

Preface

The theme of this book is to study arithmetic in a variety of number systems and to give some applications to other branches of mathematics. We are using the term 'number system' in an informal way to mean any system with well-behaved addition and multiplication. Some of the most important objects of mathematical study fit this description, for instance the systems of integers, rational numbers, real numbers, complex numbers, and integers modulo n. It may appear to be stretching the point a bit to think of polynomials as being some sort of generalized number, but the set of polynomials in one indeterminate with (for example) real-number coefficients is important and does have natural and useful operations of addition and multiplication. Over the years, mathematicians have developed these and other systems in order to tackle various problems such as solving equations. Finding the eigenvalues of a matrix involves solving a polynomial equation, and many problems in number theory are concerned with finding integer or rational number solutions to polynomial equations, and there are many other examples we could give.

It is natural to ask whether familiar properties of the arithmetic of integers extend to these more general systems. For example, it may seem to be self-evident that every integer greater than 1 is a unique product of prime numbers, but it is definitely not obvious whether or not this 'unique factorization' property holds in more general systems or even what it means. On the other hand, the importance of such questions can be illustrated by a single problem (although there are many others), namely Fermat's last theorem: If n is a positive integer greater than 2, then the nth power of a non-zero integer cannot equal the sum of two such nth powers. One of the main reasons why this famous statement remained unproved for so long is that the number systems which are used to tackle it do not all have unique factorization. Another problem which can be difficult even for integers, let alone more general number systems, is to tell whether or not a given element factorizes properly. For integers this is the problem of determining whether a given positive integer is a prime number. Questions concerning the factorization of polynomials arise in connection with the unsolvability of the quintic (there is no formula for solving polynomial equations of degree 5), the impossibility of squaring the circle, and in many other areas of mathematics.

We shall not prove Fermat's last theorem, but we shall develop some of the basic theory of such ideas as unique factorization, and will apply it to give complete solutions to some other well-known problems in number theory and geometry. Although the reader may see the theory applied later in other contexts, it is always more satisfying to see some uses immediately rather than to be told that present work will bring future benefits. In this book we shall prove Lagrange's theorem that every positive integer is the sum of four squares; Fermat's theorem that every prime number of the form $4n + 1$ is the sum of two squares; the impossibility of trisecting an arbitrary angle; Gauss's theorem that it is possible to construct a regular 17-sided polygon; and, granted that π is a transcendental number, the impossibility of squaring the circle. Such applications are attractive and were very influential in the development of the subject, but we hope that the elegance of the general theory will also appeal to the reader.

The material of this book is based on a course in commutative algebra given by the first author at the University of Bristol. In terms of students at British universities, it is suitable for undergraduates at second, third, or fourth year levels. As far as possible, we have kept each chapter short enough to be covered in three or four lectures. There are exercises at the end of each chapter, with solutions given to many of them. It is assumed that the reader is familiar with some elementary group theory (including Lagrange's theorem that the order of a subgroup of a finite group G divides the order of G), and with some basic linear algebra such as the concepts of basis and dimension in a general vector space. We have tried to keep the treatment as down-to-earth as possible, with the applications given as soon as possible and with an emphasis on worked examples with numbers in them. Homomorphisms and factor rings are not introduced until Chapter 13, but then they are put to immediate use in the construction of finite fields. For those who demand a greater challenge, there are some harder exercises and a full discussion of the construction of regular polygons.

We wish to thank the anonymous referees for many detailed and helpful comments.

Bristol   A. W. C.
Coventry   C. R. H.

August 1997

Contents

List of symbols

$\mathbb{C}$	complex numbers
J	Gaussian integers
$\mathbb{Q}$	rational numbers
$\mathbb{R}$	real numbers
$\mathbb{Z}$	integers
$\mathbb{Z}_n$	integers modulo n
$\varnothing$	empty set
$R[X]$	polynomials in X with coefficients from R
$|S|$	number of elements in S
$\partial(f)$	degree of the polynomial f
(a, b)	HCF of a and b
$o(x)$	order of element x
$[L:K]$	degree of the field extension L over K
$K(t_1, \ldots, t_r)$	the smallest field containing K and $t_1, \ldots, t_r$
$\in$	belongs to
$\notin$	does not belong to
$A \supseteq B$	A contains or equals B
$A \supsetneq B$	A strictly contains B
$A \subseteq B$	A is contained in or equals B
$A \subsetneq B$	A is strictly contained in B
$A \cap B$	intersection of A and B
$A \cup B$	union of A and B
$A \cong B$	A is isomorphic to B
$a \equiv b \,(\mathrm{mod}\, p)$	a congruent to b modulo p
$a \mid b$	a divides b
$a \nmid b$	a does not divide b
$\Rightarrow$	implies
$\Leftrightarrow$	if and only if

1 Rings

We must start by defining our terms. It is all very well to talk vaguely about generalized number systems when trying to give a non-technical feeling for the objects which will be studied in this book, but in order to make things mathematically respectable we need to pin down precisely what we are going to mean by such phrases as 'well-behaved addition'. What we are about to do when defining a ring is the analogue of setting out the rules for a game: it is necessary to have clearly formulated rules in order to know precisely what is allowed and what is not (and to settle disputes), and initially the rules may look very complicated; but with experience we get to know what is or is not permitted, and the rule-book can be put away. Similarly, the definition of a ring (which is a formalization of the idea of a generalized number system) looks complicated, but with practice we learn how to work with rings and can put aside the precise definition. So, do not be put off by the length of the definition. Also, it may seem that in this chapter we are flitting from topic to topic in a rather superficial way; this is so, but the topics we introduce briefly now will be given much more thorough development later in the book.

Many mathematical systems have well-behaved notions of addition and multiplication. In 1.1 we list some of them with their standard notation. In these examples the addition and multiplication are the obvious natural ones.

Examples 1.1

(a) The integers denoted by $\mathbb{Z}$.

(b) The complex numbers, denoted by $\mathbb{C}$.

(c) The real numbers denoted by $\mathbb{R}$.

(d) The rational numbers, denoted by $\mathbb{Q}$.

(e) The integers mod n, denoted here by $\mathbb{Z}/n\mathbb{Z}$ or by $\mathbb{Z}_n$.

(f) Polynomials in X with coefficients in the complex numbers and denoted by $\mathbb{C}[X]$.

(g) The set $R[X]$ of polynomials where the coefficients are in a more general system R with addition and multiplication.

(h) Polynomials in $X_1, \ldots, X_n$ with coefficients in $\mathbb{C}$ and denoted by $\mathbb{C}[X_1, \ldots, X_n]$.

(i) The Gaussian integers, i.e. all complex numbers of the form $a + ib$ where a and b are integers. We shall denote this by J.　■

The integers mod n and Gaussian integers are used in number theory. Polynomials in n indeterminates and with complex coefficients occur in algebraic geometry. We shall return to some of these examples later in this chapter.

All the examples in 1.1 satisfy the following definition.

Definition 1.2. A *ring* is a non-empty set R with two binary operations $+$ and $\cdot$ called addition and multiplication respectively (and with the product of x and y usually written as xy rather than $x \cdot y$) such that:

(a) The elements of R form an Abelian group under addition, i.e. for all $x, y, z \in R$ we have:

(i) $x + y \in R$ (closure),

(ii) $(x + y) + z = x + (y + z)$ (associativity),

(iii) $x + y = y + x$ (commutativity),

(iv) there is an element $0 \in R$, called the *zero element* or the additive identity, such that $x + 0 = x$ for all $x \in R$.

(v) given $x \in R$ there is an element denoted by $-x \in R$ such that $x + (-x) = 0$ (existence of additive inverses).

(b) The multiplication in R is closed, associative, commutative and R has an identity element, i.e. for all $x, y, z \in R$ we have

(i) $xy \in R$ (closure),

(ii) $xy = yx$ (commutativity),

(iii) $x(yz) = (xy)z$ (associativity),

(iv) there is an element $1 \in R$ called the (multiplicative) *identity* element or 'the 1' such that $x \cdot 1 = x$ for all $x \in R$.

(c) The distributive law holds, i.e. for all $x, y, z \in R$ we have

$$x(y + z) = xy + xz.$$

Comments 1.3

(a) From the definition it immediately follows that $0 + x = x$ for all $x \in R$.

(b) In general, $-x$ has nothing to do with being negative in the sense of being less than 0. For example if $R = \mathbb{Z}$ and $x = -2$ then $-x = 2$.

(c) The identity element 1 is sometimes called the unity or the unit element. However, these alternatives are best avoided because of possible confusion with the idea of 'a unit of a ring' which will be defined in 2.4.

(d) Unfortunately, there is no standard definition of 'ring', so when consulting a book on the subject, one should always check which form of the definition is being used. Some authors omit the commutativity law for multiplication and/or the existence of the element 1. What we are calling a ring here is sometimes called a commutative ring with identity.

(e) There is an irritating minor niggle as to whether one should allow the two special elements 0 and 1 of a ring to be equal. Some authors insist that these should be distinct but we do not. According to our definition a ring may consist of the single element 0 with $0 + 0 = 0$ and $0 \cdot 0 = 0$. In this case 0 is the multiplicative identity element, i.e. $0 = 1$. In fact this is the only case in which $0 = 1$ (see Question 2, Exercises 1).

(f) The zero element 0 of a ring R is unique, because it is the identity element of the group consisting of the elements of R under addition. Also given $x \in R$ its additive inverse $-x$ is unique. To show that the identity element of R is unique, suppose that e and f are both multiplicative identity elements of R. As e is an identity element we have $ef = f$. However, f is also an identity element. Hence $ef = e$. Therefore $e = f$. We are not able to use group theory to prove this because in general the elements of a ring do not form a group under multiplication, even if the element 0 is deleted. For example, if $R = \mathbb{Z}$ and $x = 2$, there is no element $y \in R$ such that $xy = 1$ (i.e. 2 does not have a multiplicative inverse in $\mathbb{Z}$).

(g) A ring is sometimes displayed as the triple $(R, +, \cdot)$ to show that it consists of the set R with the operations $+$ and $\cdot$. When considering a ring R it is of course necessary to know what addition and multiplication are defined on R. These may however be left unspecified if they are obvious or clear from the context.

(h) We shall write $x - y$ to mean $x + (-y)$.

(i) The dot is usually omitted from the notation for products, but is sometimes re-introduced for greater clarity.

Standard properties follow easily and in 1.4 we list some of them.

Proposition 1.4. *Let R be a ring. Then*

(i) $0 \cdot x = 0$ *for all* $x \in R$.

(ii) $-(-x) = x$ *for all* $x \in R$.

(iii) $x(-y) = (-x)y = -xy$ *for all* $x, y \in R$.

(iv) $(-x)(-y) = xy$ *for all* $x, y \in R$.

Proof. We shall prove (i), (ii) and part of (iii). The proof of (iv) is similar.

(i) Let $x \in R$. Set $y = 0 \cdot x$. We have $0 = 0 + 0$. Hence $0 \cdot x = (0 + 0) \cdot x = 0 \cdot x + 0 \cdot x$. So $y = y + y$. Therefore $0 = y - y = (y + y) - y = y + (y - y) = y + 0 = y$.

(ii) We have $(-x) + x = 0$. So $x = -(-x)$ by the uniqueness of the negative for $-x$.

(iii) $xy + (-x)y = [x + (-x)]y = 0y = 0$ by (i). Hence $(-x)y = -xy$ by the uniqueness of $-xy$. ■

For x an element of a ring and n a positive integer x^n is defined to be $x \cdot x \cdots \cdot x$ (n times). It is easily seen that for positive integers m and n, we have $x^m x^n = x^{m+n}$ and $(x^m)^n = x^{mn}$. It is conventional to take $x^0 = 1$.

Non-examples 1.5. The following are not rings according to *our* definition

(a) The set of even integers has all the properties of a ring except that it has no multiplicative identity element. Note that the integer 1 is not an identity element in this case because it does not belong to the set.

(b) The set of all 2×2 matrices with entries in the complex numbers has all the properties of a ring except that the multiplication of two elements is not always commutative.
    For example

$$\begin{bmatrix} 1 & 0 \\ 0 & 0 \end{bmatrix}\begin{bmatrix} 0 & 1 \\ 0 & 0 \end{bmatrix} = \begin{bmatrix} 0 & 1 \\ 0 & 0 \end{bmatrix} \quad \text{but} \quad \begin{bmatrix} 0 & 1 \\ 0 & 0 \end{bmatrix}\begin{bmatrix} 1 & 0 \\ 0 & 0 \end{bmatrix} = \begin{bmatrix} 0 & 0 \\ 0 & 0 \end{bmatrix}. \quad ■$$

We next revisit some of the examples listed in 1.1 and make some further comments. The verification that these examples satisfy our definition of a ring is straightforward but tedious and we leave it as an exercise.

Notation 1.6. The symbol $a \mid b$ will stand for a divides b. Thus $5 \mid 60$, but $7 \nmid 60$.

Definition 1.7. Let $a, b, n \in \mathbb{Z}$ with n positive. Then a is said to be *congruent* to b (mod n) if $n \mid a - b$. This is usually denoted by $a \equiv b$ (mod n).

Example 1.8. Let n be a positive integer. Let the symbol $\mathbb{Z}/n\mathbb{Z}$ denote the set of integers (mod n). This set can be thought of as the integers with '$n = 0$'. Usually the most convenient way of representing the elements of $\mathbb{Z}/n\mathbb{Z}$ is to use the symbols $0, 1, 2, \ldots, n - 1$ (with $n = 0$). These are the distinct congruence classes (mod n). The elements are added and multiplied in the usual way and, if necessary, the answer is reduced (mod n) by throwing away multiples of n to get it in the range from 0 to $n - 1$. That $\mathbb{Z}/n\mathbb{Z}$ is a ring can be checked directly using Question 1, Exercises 1. However, this is more satisfactorily done by interpreting $\mathbb{Z}/n\mathbb{Z}$ as a factor ring of $\mathbb{Z}$ by the ideal $n\mathbb{Z}$ (13.35). Equality in $\mathbb{Z}/n\mathbb{Z}$ can be written as congruence (mod n). However, we shall also from time to time use the ordinary equality sign $=$ accompanied by the words 'working in $\mathbb{Z}/n\mathbb{Z}$' or a similar expression. From this point of view, integers a and b are equal elements of $\mathbb{Z}/n\mathbb{Z}$ if and only if in $\mathbb{Z}$ the integer $a - b$ is divisible by n. For example, (mod 5) we have $16 \equiv 1$ so that $4^5 \equiv 16 \cdot 16 \cdot 4 \equiv 1 \cdot 1 \cdot 4 \equiv 4$. An alternative display which is often employed when n is odd is to use both 'positive' and 'negative' symbols distributed symmetrically about 0. For example, the elements of $\mathbb{Z}/5\mathbb{Z}$ can be represented by the symbols -2, -1, 0, 1, 2. This has the advantage of reducing the 'size' of the numbers and simplifying calculations. For example, (mod 5) we have $4^5 \equiv (-1)^5 \equiv -1 \equiv 4$. If we have a doubt concerning the validity of some calculation in $\mathbb{Z}/n\mathbb{Z}$, we can always translate it to a problem in $\mathbb{Z}$ by the following procedure: '$x^2 = 1$ in $\mathbb{Z}/n\mathbb{Z}$' is equivalent to 'in $\mathbb{Z}$ we have $x^2 = 1 + kn$ for some integer k'. For example, suppose that we have $2x = 2$ in $\mathbb{Z}/6\mathbb{Z}$. Is it valid to deduce that $x = 1$? The safest method is to say that in $\mathbb{Z}$ we have $2x = 2 + 6k$ for some integer k. Hence $x = 1 + 3k$ and it follows that in $\mathbb{Z}/6\mathbb{Z}$ we can have $x = 1$ or $x = 4$.

We started by saying 'Let n be a positive integer'. What happens if $n = 1$? Recall that '$a \equiv b$ (mod n)' means that $a - b$ is divisible by n in $\mathbb{Z}$. If $n = 1$ then $a - b$ is divisible by n for all integers a and b. In particular, $a - 0$ is divisible by 1 for all integers a so that $a \equiv 0$ (mod 1) for all a. Thus $\mathbb{Z}/1\mathbb{Z}$ consists of the single element 0, and $\mathbb{Z}/1\mathbb{Z}$ is a ring in which $0 = 1$. ∎

Example 1.9. Let R be any ring and let $R[X]$ denote the ring of all polynomials in X with coefficients in R. By definition, a polynomial has only a finite number of non-zero coefficients. It is easy to verify that $R[X]$ is a ring with respect to the usual addition and multiplication of polynomials. Elements of R considered as elements of $R[X]$ are sometimes referred to as *constants*. The correct name for X in this context is 'indeterminate', not 'variable'. The important thing about a polynomial is its coefficients and not its indeterminate X, in the sense that the powers of X are merely coat-hangers on which we drape the coefficients. Two polynomials are *defined* to be equal if they have identical coefficients. Thus the equality of two polynomials is determined by their coefficients and not by the values they take as functions. Thus, for example, if we take R to be the ring $\mathbb{Z}/5\mathbb{Z}$ whose elements are denoted as integers (mod 5), we find that $0^5 \equiv 0$, $1^5 \equiv 1$, $2^5 \equiv 32 \equiv 2$, $3^5 \equiv 3 \cdot 81 \equiv 3 \cdot 1 \equiv 3$, $4^5 \equiv 16 \cdot 16 \cdot 4 \equiv 1 \cdot 1 \cdot 4 \equiv 4$. Thus the function which sends a to a^5 for all $a \in R$ is the same as the function which sends a to a. However, X^5 and X are distinct elements of $R[X]$.

The *degree* $\partial(f)$ of a polynomial $f(X)$ is defined to be the largest integer n such that the coefficient of X^n in $f(X)$ is non-zero. Thus in $\mathbb{Z}[X]$ we have

$$\partial(2X^5 + 5X^3 + 2X^2 - 7) = 5.$$

There is a problem of what we mean by $\partial(f)$ when $f(X)$ is the zero polynomial, i.e. the polynomial all of whose coefficients are 0. The difficulty arises in trying to make the familiar rule for degrees of products $\partial(fg) = \partial(f) + \partial(g)$ valid even when $f(X)$ is the zero polynomial. For instance, in $\mathbb{Z}[X]$ if we take $f(X) = 0$ and $g(X) = X^2 + 1$ we would have to have $\partial(0) = \partial(0) + 2$. For this reason some authors define $\partial(0)$ to be $-\infty$. Others define it to be 0 so that all constant polynomials have degree 0. Following a third convention, we shall not give a meaning to $\partial(0)$. Therefore we shall always have to treat the zero polynomial separately when the degree of a polynomial is being considered.

If $f(X)$ is a polynomial, we may from time to time abbreviate the symbols to f for convenience if no confusion is likely to arise.   ∎

Definition 1.10. Let S be a non-empty subset of a ring R. Then S is called a *subring* of R if S is also a ring with respect to the operations given in R.

Proposition 1.11. *Let S be a subset of a ring R. In order to show that S is a subring of R it is enough to verify the following:*

(1) *S has a multiplicative identity element (which need not be the same as that of R—see 1.14).*

(2) *If $x, y \in S$ then $x - y \in S$ and $xy \in S$.*

Proof. Let $x \in S$. Then $0 = x - x \in S$ by (2). We also have $0 - x \in S$ since $0 \in S$ and $x \in S$. Thus $x \in S$ implies that $-x \in S$. Now let $x \in S$ and $y \in S$. From above we know that S contains both x and $-y$. Hence by (2) we have $x - (-y) \in S$ and so $x + y \in S$. Thus S is closed under addition. We are given in (2) that S is closed under multiplication. The zero element of R acts as the zero element of S and we are assuming in (1) that S has an identity element. The remaining axioms for a ring, such as the associative and the distributive laws, hold in S because they hold in R.   ∎

Example 1.12. Let J be the set of *Gaussian integers*, i.e. the set of all complex numbers of the form $a + ib$ where $a, b \in \mathbb{Z}$ and $i^2 = -1$. Since J is a subset of the ring $\mathbb{C}$ of complex numbers, we can apply 1.11 to show that J is a ring. Clearly $1 \in J$ and 1 is obviously an identity element for J. Let $a + ib$ and $c + id \in J$. Then $a, b, c, d \in \mathbb{Z}$. We have

$$(a + ib) - (c + id) = (a - c) + i(b - d) \quad \text{with} \quad a - c, b - d \in \mathbb{Z}.$$

Therefore $(a + ib) - (c + id) \in J$. Also

$$(a + ib)(c + id) = (ac - bd) + i(ad + bc) \in J$$

since $ac - bd$ and $ad + bc \in \mathbb{Z}$. Thus J is a subring of $\mathbb{C}$.   ∎

One of the reasons why the Gaussian integers are useful in number theory is that they have an associated *norm function*. This function N is defined by $N(a + ib) = a^2 + b^2$ for all $a, b \in \mathbb{Z}$. The set of values taken by the function N consists precisely of those non-negative integers which can be expressed as a sum of two squares $a^2 + b^2$. This means that J and N can be used in questions concerning sums of squares (see 4.1 and 6.15). By noting that $N(a + ib)$ is the square of the absolute value (modulus) of the complex number $a + ib$, it is easy to show that $N(xy) = N(x)N(y)$ for all $x, y \in J$. This multiplicative property of N enables us to change questions about the multiplication in J into questions about multiplication in $\mathbb{Z}$ (cf. 6.16).

Example 1.13. Let S be the set of non-negative elements of $\mathbb{Z}$. Then S is not a subring of $\mathbb{Z}$ because, for example, the element $2 \in S$ has no additive inverse, i.e. there is no element $x \in S$ such that $2 + x = 0$. However, $0, 1 \in S$ and S is closed under addition as well as multiplication. This shows why we ask for closure under *subtraction* in 1.11(2).   ∎

Example 1.14. Let R be the set of all ordered pairs (x, y) with $x, y \in \mathbb{Z}$. We define addition and multiplication in R by setting $(a, b) + (c, d) = (a + c, b + d)$ and $(a, b)(c, d) = (ac, bd)$ for all $a, b, c, d \in \mathbb{Z}$. It is straightforward to show that R is a ring with respect to these operations. The zero and identity elements of R are $(0,0)$ and $(1,1)$ respectively. Set $S = \{(x,0) \mid x \in \mathbb{Z}\}$. Then S is a subset of R which satisfies (1) and (2) of 1.11, the identity of S being the element $(1,0)$. Thus S is a ring in its own right with respect to the operations which it inherits from R. However, the identity element $(1,0)$ of S is not the same as the identity element $(1,1)$ of R.   ■

Exercises 1

1. Let a, b, c, d and n be integers with n positive such that $a \equiv b$ and $c \equiv d$ (mod n). Show that $a + c \equiv b + d$ and $ac \equiv bd$ (mod n).

2. Let R be a ring where, as usual, 0 and 1 denote the zero and the identity, respectively. Show that $0 = 1$ in R if and only if 0 is the only element of R.

3. Let S be the set of all rational numbers a/b where a and b are integers with b odd. Show that S is a ring.

4. Show by an example that the degree formula $\partial(fg) = \partial(f) + \partial(g)$ for the product of polynomials fails if the coefficients are in $\mathbb{Z}/4\mathbb{Z}$. What property of $\mathbb{Z}/4\mathbb{Z}$ causes this failure?

5. Let R be the set of all integers with the usual notation for addition and multiplication. Given $a, b \in R$, define the new sum and the new product of a and b to be $a + b + 1$ and $ab + a + b$ respectively. Show that R is a ring with respect to the new operations. Is R a subring of $\mathbb{Q}$ under these new operations?

6. Let R be the set of all subsets of a fixed non-empty set S. Let $A, B \in R$. Define $A + B$ to be the set of all elements of S which are in A or B but not in both. Define AB to be $A \cap B$. Convince yourselves, using Venn diagrams if you wish, that R is a ring with respect to these operations. Note that $x^2 = x$ for all $x \in R$. A ring with this property is called a *Boolean ring*. What happens if S is empty?

7. Let d be a fixed integer with no repeated prime factors and let R be the set of all complex numbers of the form $a + b\sqrt{d}$ for all $a, b \in \mathbb{Z}$. Prove that R is a ring. Set $N(a + b\sqrt{d}) = a^2 - db^2$ for all $a, b \in \mathbb{Z}$. Show that $N(xy) = N(x)N(y)$ for all $x, y \in R$.

8. Let w be a complex number such that $w^3 = 1$ and $w \neq 1$. Show that $w^3 - 1 = (w - 1)(w^2 + w + 1)$ and deduce that $w^2 + w + 1 = 0$. Let R

be the set of all complex numbers of the form $a + bw$ for $a, b \in \mathbb{Z}$. Prove that R is a ring. Show that if x and y are integers then

$$x^3 + y^3 = (x + y)(x + wy)(x + w^2y).$$

This linear factorization is the reason why the ring R is used to deal with Fermat's equation $X^n + Y^n = Z^n$ when $n = 3$.

9. Find all the solutions in $\mathbb{Z}/6\mathbb{Z}$ of the following equations:

(i) $4x = 2$; (ii) $4x = 3$; (iii) $x^2 + 3x + 2 = 0$.

Why is it not possible just to use the familiar formula for solving quadratic equations?

2 Euclidean rings

The theme of this chapter is the study of those very special rings which have a division algorithm. For integers this states that any integer a can be divided by any positive integer b in such a way as to leave a small remainder (i.e. a remainder in the range from 0 to $b - 1$). For instance, we can divide 100 by 7 with remainder 2 because $100 = 14 \cdot 7 + 2$ (i.e. 7 into 100 goes 14 times with 2 left over). Also we can divide 140 by 7 with remainder 0 because $140 = 20 \cdot 7 + 0$. The general sort of notation which occurs is $a = qb + r$ where q and r are the so-called quotient and remainder when a is divided by b. Although the quotient and remainder are unique in the context of integers, we shall not emphasize this because they are not in general unique in some of the other contexts which we shall consider.

The division algorithm for integers is enormously useful, and so it makes sense to investigate whether something similar and equally useful works, say, for Gaussian integers. When it works (as indeed it does for Gaussian integers), it can be used as the foundation for a successful extension, to the rings in question, of arithmetic ideas such as prime numbers and highest common factors. It is then possible to use these new arithmetics to solve problems in classical number theory and some other branches of mathematics. We shall give some of these applications in later chapters.

We begin by proving the very important division algorithm for integers.

Theorem 2.1 (The division algorithm for $\mathbb{Z}$). *Let $a, b \in \mathbb{Z}$ with $b \neq 0$. Then there exist $q, r \in \mathbb{Z}$ with $0 \leq r < |b|$ such that $a = bq + r$.*

Proof. Let x be the largest integer such that $x \leq a/|b|$. Set $r = a - |b| x$. Then $r \in \mathbb{Z}$ and $r \geq 0$. Also we must have $x + 1 > a/|b|$ and so $|b| x + |b| > a$. Therefore $|b| > a - |b| x$ and hence $|b| > r$.

Now set $q = \begin{cases} x & \text{if } b > 0 \\ -x & \text{if } b < 0. \end{cases}$

Then clearly $a = bq + r$ with $0 \leq r < |b|$. ∎

The division algorithm plays a fundamental role in the theory of integers. It means that an integer can be divided by a non-zero integer either exactly or with a 'smaller' remainder. A similar property plays a crucial part in several other important rings. They are called Euclidean rings and include the integers, the Gaussian integers, and polynomial rings with rational, real or complex coefficients. Any such ring also has the unique factorization property for its elements. We introduce some of these ideas here. They will be developed further in later chapters.

Definition 2.2. An *integral domain* (sometimes just called a *domain*) is a ring with the property that $1 \neq 0$ and if a and b are non-zero elements of R then $ab \neq 0$.

Note that 2.2 is phrased so that the ring consisting of the single element 0 is, by definition, not an integral domain. The more important part of the definition of an integral domain is that the product of non-zero elements is always non-zero. Of course, every integral domain is a ring but not every ring is an integral domain. For example, in the ring $\mathbb{Z}/6\mathbb{Z}$ we have $2 \cdot 3 = 0$ but $2 \neq 0$ and $3 \neq 0$. All the rings in 1.1 are integral domains except for $\mathbb{Z}/n\mathbb{Z}$ when n is not prime (see 5.7) and the ring $R[X]$ when R itself is not an integral domain.

Proposition 2.3. *Let R be an integral domain and $a, b, c \in R$. Suppose that $ab = ac$. If $a \neq 0$ then $b = c$.*

Proof. $ab = ac$ and so $a(b - c) = 0$. If $a \neq 0$ we have $b - c = 0$ and so $b = c$. ∎

Thus in an integral domain non-zero elements can be cancelled.

Definition 2.4. (i) An element x of a ring R is said to be a *unit* of R if there is an element y of R such that $xy = 1$. In these circumstances we write $y = x^{-1}$ and say that y is the *inverse* of x (in fact, such an element y is uniquely determined by x).

(ii) A ring R is called a *field* if R has at least two elements and every non-zero element of R has an inverse.

We shall study fields in greater detail later, but for the purposes of this chapter it is enough to note that every field is an integral domain. However, an integral domain need not be a field as the example $\mathbb{Z}$ shows. The most familiar examples of fields are $\mathbb{Q}$, $\mathbb{R}$ and $\mathbb{C}$, and at this stage it may help to think of these particular examples whenever the word 'field' occurs.

Examples 2.5

(a) The units of $\mathbb{Z}$ are 1 and -1, i.e. the only integers x such that $xy = 1$ for some *integer y* are $x = \pm 1$.

(b) $\mathbb{Q}$, $\mathbb{R}$, $\mathbb{C}$ and $\mathbb{Z}/p\mathbb{Z}$ for prime integers p, are examples of fields (see Chapter 5).

(c) If F is a field, the units of $F[X]$ are the non-zero constant polynomials, i.e. the non-zero elements of F.

(d) The units of $\mathbb{Z}[X]$ are ± 1.

(e) $1 + 2X$ is a unit of the ring $(\mathbb{Z}/4\mathbb{Z})[X]$ because $(1 + 2X)^2 = 1$.

(f) i is a unit of the ring of Gaussian integers because $i \cdot (-i) = 1$.

(g) Let A be the ring of real numbers of the form $a + b\sqrt{2}$ with $a, b \in \mathbb{Z}$. Set $u = 1 + \sqrt{2}$. Then u is a unit because $(1 + \sqrt{2})(-1 + \sqrt{2}) = 1$. Since $u \in \mathbb{R}$ and $u > 1$, the numbers $u, u^2, u^3, \ldots$ are distinct and it is easy to show that they are units of A. Hence A has infinitely many units. It can be shown that an element of A is a unit if and only if it is of the form u^n or $-u^n$ for some $n \in \mathbb{Z}$.   ∎

It can be very difficult to find all the units of a ring. Example 2.5(g) indicates this. Nevertheless in questions of factorization it is vital to know the units (or at least enough about them) because these are the trivial factors as explained in Chapter 6. For example, we do not regard $6 = 2 \cdot 3 = (-2) \cdot (-3)$ as violating the unique factorization property for integers, because changing 2 and 3 to -2 and -3 involves a factor -1 and -1 is trivial, i.e. -1 is a unit of $\mathbb{Z}$.

Definition 2.6. An integral domain R is said to be a *Euclidean domain* or *Euclidean ring* if there is a function N from the non-zero elements of R to the non-negative integers such that:

(i) Given $a, b \in R$ with $b \neq 0$ there are elements q and $r \in R$ (called respectively *quotient* and *remainder*) such that $a = bq + r$ where either $r = 0$ or $N(r) < N(b)$ (the *division algorithm*).

(ii) For all non-zero elements a and $b \in R$, we have $N(a) \leq N(ab)$.

Comments 2.7

(a) The reason for (ii) in the above definition is technical and is to do with recognizing units (see below).

(b) Note that in 2.6 we do not need $N(0)$ to be defined.

(c) Note also that in 2.6(i) we do not require the quotient and the remainder, q and r, to be unique.

(d) If there were non-zero elements a and b of R with $ab = 0$, then 2.6(ii) would not make sense. Thus Definition 2.6 is applied only to integral domains and the terms 'Euclidean ring' and 'Euclidean domain' mean the same thing.

(e) The proofs of general results about Euclidean rings are usually similar to the proofs of the corresponding results for $\mathbb{Z}$; but the advantage of proving them in general is that the results then apply to other examples such as the Gaussian integers (see below).

In number theory, showing that a certain ring is Euclidean is often only a means to the end of establishing the unique factorization property for elements.

(f) The function N in 2.6 is called the *norm*, the *Euclidean norm* or the *Euclidean valuation*.

Example 2.8. (a) The ring $\mathbb{Z}$ is Euclidean with respect to the norm $N(z) = |z|$ for $z \in \mathbb{Z}$. The division algorithm was proved in 2.1. For non-zero $a, b \in \mathbb{Z}$ we have $1 \le |b|$ and so $|a| \le |a| \, |b| = |ab|$. Thus $N(a) \le N(ab)$.

(b) It is possible to have more than one norm on a Euclidean domain. We shall show that $\mathbb{Z}$ is also Euclidean with respect to the norm N given by $N(z) = z^2$ for all $z \in \mathbb{Z}$. Given $a, b \in \mathbb{Z}$ with $b \ne 0$, we know by 2.1 that $a = bq + r$ for some $q, r \in \mathbb{Z}$ with $0 \le r < |b|$. Clearly then we have either $r = 0$ or $r^2 < b^2$, i.e. either $r = 0$ or $N(r) < N(b)$. Also $N(a) = a^2 \le a^2 b^2 = (ab)^2 = N(ab)$. Thus $\mathbb{Z}$ is Euclidean with respect to N. Here the quotient and the remainder are not unique. Take $a = 3$ and $b = 2$. We have $a = bq + r$ where either $q = r = 1$ or $q = 2$ and $r = -1$. In both cases $N(r) = 1 < 4 = N(b)$.

(c) We cannot use the function N given by $N(z) = z$ because this takes negative values. ∎

Theorem 2.9. *Let F be a field and $R = F[X]$ the ring of polynomials with coefficients in F. Then R is a Euclidean domain with respect to the norm defined by $N(r) = \partial(r)$ for non-zero $r \in R$.*

Proof. Let $f(X), g(X) \in F[X]$ with $g \ne 0$. In order to prove the division algorithm we must show that

$$f(X) = g(X)q(X) + r(X)$$

for some $q(X), r(X) \in F[X]$ where either $r(X) = 0$ or $\partial(r) < \partial(g)$. If $f(X) = 0$ or $\partial(f) < \partial(g)$ we can simply take $q(X) = 0$ and $r(X) = f(X)$. Therefore from now on we shall suppose that $\partial(f) \ge \partial(g)$. Set $n = \partial(f)$ and $k = \partial(g)$. Let $a_n X^n$ and $b_k X^k$ be the leading terms of $f(X)$ and $g(X)$ respectively, where a_n and b_k are non-zero elements of F.

Set $q_1(X) = (a_n/b_k)X^{n-k}$. Note that $q_1(X) \in F[X]$ since we are assuming $n \ge k$. The leading term of $g(X)q_1(X)$ is $b_k X^k(a_n/b_k)X^{n-k} = a_n X^n$. Thus the polynomials $f(X)$ and $g(X)q_1(X)$ have the same leading term so that $f(X) - g(X)q_1(X)$ is either the zero polynomial or it has degree less than n. Now we repeat the process with $f(X) - g(X)q_1(X)$ in place of $f(X)$. After a finite number of steps we obtain an element $q(X) \in F[X]$

such that $f(X) - g(X)q(X)$ is either the zero polynomial or has degree
less than $k = \partial(g)$. Set $r(X) = f(X) - g(X)q(X)$. This proves the division
algorithm.

Finally let $a(X)$ and $b(X)$ be any non-zero elements of $F[X]$. Then
$\partial(a) \leq \partial(a) + \partial(b) = \partial(ab)$.   ■

We shall now show that the ring J of Gaussian integers is Euclidean.
First we prove a lemma.

Lemma 2.10.   *Let $a, b \in J$ with $b \neq 0$. Then there exists $q \in J$ such that
$|a/b - q| \leq 1/2$.*

Proof. $a/b \in \mathbb{C}$. So let $a/b = u + iv$ with $u, v \in \mathbb{R}$. Let c be an integer
as close to u as possible. Similarly, choose d to be an integer as close
as possible to v. Then $|u - c| \leq 1/2$ and so $|u - c|^2 \leq 1/4$. Similarly,
$|v - d|^2 \leq 1/4$. Let $q = c + id$. Then

$$|a/b - q|^2 = |u + iv - c - id|^2$$
$$= (u - c)^2 + (v - d)^2 \leq 1/4 + 1/4 = 1/2.$$ ■

Theorem 2.11.   *J is a Euclidean domain.*

Proof.   We define a norm on J as follows. Let $r \in J$. We have $r = x + iy$
for some $x, y \in \mathbb{Z}$. Set $N(r) = x^2 + y^2$. Clearly $N(r)$ is a non-negative
integer and $N(r) = 0$ if and only if $r = 0$.

Let $a, b \in J$ with $b \neq 0$. By 2.10 there exists $q \in J$ with $|a/b - q|^2 \leq 1/2$.
Therefore

$$|a - bq|^2 \leq \frac{|b|^2}{2} < |b|^2.$$

Put $a - bq = r$. We have $a = bq + r$ with $N(r) < N(b)$ (which holds in this
case for $r = 0$ also).

Now let s, t be non-zero elements of J. We must show that $N(st) \geq N(s)$.
However $N(st) = N(s)N(t)$ (see the paragraph after 1.12). Also $N(s), N(t)$
are positive integers. Hence $N(st) \geq N(s)$.   ■

Proposition 2.12.   *Let R be a Euclidean ring with respect to the norm N.*

(i) *$N(1)$ is the minimum value taken by $N(a)$ as a ranges over the non-zero
    elements of R.*

(ii) *If u is a non-zero element of R then u is a unit if and only if $N(u) = N(1)$.*

Proof. (i) Let b be any non-zero element of R. Taking $a = 1$ in 2.6(ii) we
have $N(1 \cdot b) \geq N(1)$. Thus $N(b) \geq N(1)$.

(ii) Firstly let u be a unit of R. Then $uv = 1$ for some $v \in R$. By 2.6(ii)
we have $N(uv) \geq N(u)$. Thus $N(1) \geq N(u)$ and by (i) it follows that
$N(u) = N(1)$.

Conversely, suppose that u is a non-zero element of R with $N(u) = N(1)$.
By the division algorithm we have $1 = qu + r$ for some $q, r \in R$ where
either $r = 0$ or $N(r) < N(u)$. However, $N(u) = N(1)$ and by (i) there is no
non-zero element $r \in R$ with $N(r) < N(1)$. Therefore $r = 0$. Thus $qu = 1$
and u is a unit with inverse q.   ∎

The above proposition is useful for finding the units in a Euclidean
domain.

Example 2.13. Let $\mathbb{Z}$ and N be as in 2.8(b) and let $u \in \mathbb{Z}$. Then u is a
unit of $\mathbb{Z}$ if and only if $N(u) = N(1)$, i.e. $u^2 = 1$. Thus u is a unit if and only
if $u = 1$ or $u = -1$.   ∎

Example 2.14. Let J be the ring of Gaussian integers and let N be as in
2.11. Let $u = a + ib$ with $a, b \in \mathbb{Z}$. Then u is a unit of $J \Leftrightarrow N(u) = N(1) \Leftrightarrow$
$a^2 + b^2 = 1 \Leftrightarrow$ either $a^2 = 1$ and $b^2 = 0$ or $a^2 = 0$ and $b^2 = 1$. Thus the
units of J are $1, -1, i, -i$.   ∎

Example 2.15. Let R be the ring of all real numbers of the form $a + b\sqrt{2}$
for $a, b \in \mathbb{Z}$. It is an exercise (Question 9, Exercises 2) to show that R is
Euclidean with respect to the function N given by $N(a + b\sqrt{2}) = |a^2 - 2b^2|$.
Hence $a + b\sqrt{2}$ is a unit of $R \Leftrightarrow N(a + b\sqrt{2}) = N(1) \Leftrightarrow |a^2 - 2b^2| = 1 \Leftrightarrow$
$a^2 - 2b^2 = \pm 1$.

This last equation has many solutions; e.g. $a = \pm 1$ and $b = 0$; $a = \pm 3$
and $b = \pm 2$, $a = \pm 7$ and $b = 5$. Equations of this form are studied in
number theory under the heading 'Pell's equation'.   ∎

Exercises 2

1. Show that the integers q and r occurring in 2.1 are unique.

2. Prove that $R[X]$ is an integral domain if and only if R is an integral
   domain.

3. Let R be an integral domain. Let f, g be non-zero elements of $R[X]$.
   Show that $\partial(fg) = \partial(f) + \partial(g)$.

4. Show that the units of a ring form a group with respect to multiplication.

5. Let d be an integer and let R be the ring of all $a + b\sqrt{d}$ with $a, b \in \mathbb{Z}$. Find some non-trivial units of R in each of the following cases: (i) $d = 3$ (ii) $d = 5$ (iii) $d = 6$ (iv) $d = 7$.

6. Let F be a field and set $N(x) = 1$ for all non-zero elements $x \in F$. Show that F is a Euclidean ring with respect to N.

7. Show that $\mathbb{Z}[X]$ is not Euclidean with respect to the function N given by $N(f(X)) = \partial(f)$. (See also 14.3 and 14.5).

8. Give an example to show that the quotient and remainder are not unique for the division algorithm in the Gaussian integers.

9. Let d be an integer and let R be the ring of all numbers $a + b\sqrt{d}$ for $a, b \in \mathbb{Z}$. Let $N(a + b\sqrt{d}) = |a^2 - b^2 d|$ for all $a, b \in \mathbb{Z}$ (see Question 7, Exercises 1). Modify the proof of 2.11 to show that R is Euclidean with respect to N if $d = 2, 3$, or -2.

10. Fill in the details in the following proof. Set $w = (1 + \sqrt{(-3)})/2$. Then $w^2 = w - 1$. Let A be the set of all complex numbers of the form $a + bw$ for $a, b \in \mathbb{Z}$. Then A is a subring of $\mathbb{C}$. For each $r \in A$ let $N(r) = |r|^2$. Then $N(r)$ is a non-negative integer for each $r \in A$. Every complex number can be written in the form $u + vw$ for some $u, v \in \mathbb{R}$. The method of taking the nearest integer can be used as in 2.10 to show that A satisfies the division algorithm with respect to N. It then follows easily that A is Euclidean. The units of A are $1, -1$, $(1 + \sqrt{(-3)})/2$, $(1 - \sqrt{(-3)})/2$, $(-1 + \sqrt{(-3)})/2$ and $(-1 - \sqrt{(-3)})/2$.

3 Highest common factor

In this chapter we shall consider divisibility, common factors, and the existence of highest common factors in arbitrary integral domains. From this work we shall deduce properties of integers which, as well as being of interest in their own right, will be needed in later chapters. Although our main interest in the short term is the ring of integers, we shall prove some of the results in a more general form which will be useful for later applications. Roughly speaking, the sort of results which we are interested in for integers can usually be extended to arbitrary Euclidean domains with only minor modifications to the proofs.

One particular result which we shall prove for integers (Theorem 3.16) is usually known as Fermat's little theorem. This is named after Pierre de Fermat (1601–65), and is extremely useful in number theory even though it is relatively easy to prove. It is known as the little theorem to distinguish it from the much more famous result known as Fermat's last theorem (which was not proved to be true until recently): if n is a positive integer which is 3 or larger, then an nth power cannot be separated into two nth powers (i.e. there are no non-zero integers x, y, z such that $x^n + y^n = z^n$).

Definition 3.1. Let a, b be elements of an integral domain R. Then a is said to *divide* b in R if $b = ac$ for some $c \in R$. As with integers we shall write $a \mid b$. If $a \mid b$ then a is said to be a *factor* of b.

Comments 3.2. Let R be an integral domain. The following statements follow easily from the definition.

(a) Every element of R divides 0.

(b) The only element of R which is divisible by 0 is 0.

(c) The units of R divide every element of R.

(d) The units of R are precisely the elements of R which divide 1.

(e) If a, b, c are elements of R such that $a \mid b$ and $b \mid c$ then $a \mid c$.

(f) Let a, b, c be elements of R such that $a \mid b$ and $a \mid c$. Then $a \mid (bx + cy)$ for all $x, y \in R$.

(g) 3 does not divide 4 in $\mathbb{Z}$ since there is no $x \in \mathbb{Z}$ such that $3x = 4$. However, 3 divides 4 in $\mathbb{Q}$ since $3x = 4$ with $x = 4/3$.

Definition 3.3.   Let a and b be elements of an integral domain R. An element h of R is said to be a *highest common factor* (HCF) or a *greatest common divisor* of a and b if the following two conditions hold:

(i) $h \mid a$ and $h \mid b$.

(ii) If $d \in R$ and $d \mid a$ and $d \mid b$ then $d \mid h$.

We write $h = (a, b)$ if h is an HCF of a and b.

Comments 3.4.   Let a and b be elements of an integral domain R.

(a) We are not claiming in 3.3 that a and b always have an HCF.

(b) The definition of HCF given in 3.3 is expressed purely in terms of divisibility and not in terms of size or 'largest' common divisor.

(c) An HCF of a and b, if it exists, is not usually unique. For example, if $R = \mathbb{Z}$, $a = 4$ and $b = -18$, then both 2 and -2 are HCFs of a and b according to 3.3. In general, if h is an HCF of a and b, so also is uh for every unit $u \in R$, and every HCF of a and b is of the form uh for some unit $u \in R$ (see Question 2, Exercises 3).

(d) Even when the HCF is not unique, the term 'the HCF' is often used to refer to any one of the HCFs.

In the next proposition and later on we shall use the property of integers that every non-empty set of non-negative integers has a smallest element.

Proposition 3.5.   *Let a and b be non-zero elements of a Euclidean ring R. Then a and b have an HCF. If h is the HCF then $h = sa + tb$ for some $s, t \in R$.*

Proof.   Let $I = \{xa + yb \mid x, y \in R\}$. Then I contains both a and b because, for example, $a = 1a + 0b$. In particular, it follows that I contains some non-zero elements. Suppose that R is Euclidean with respect to the function N. As c ranges over the non-zero elements of I, the corresponding values $N(c)$ form a non-empty set of non-negative integers. Therefore there is a non-zero element $h \in I$ such that $N(h)$ is the minimum value of $N(c)$ for non-zero elements $c \in I$. Since $h \in I$, we have $h = sa + tb$ for some $s, t \in R$. It remains to show that h is an HCF of a and b. Clearly any common factor of a and b also divides h. We must show that h divides both a and b. By the division algorithm we have $a = qh + r$ for some $q, r \in R$ where either $r = 0$ or $N(r) < N(h)$. However, $r = a - qh = a - q(sa + tb) = (1 - qs)a - qtb$, so that $r \in I$. By the minimality of $N(h)$ we cannot have $r \neq 0$. Therefore $r = 0$ and $a = qh$. Thus $h \mid a$. Similarly, $h \mid b$.   ∎

It is not always true, even when a ring R is a UFD (see 6.4) that the HCF of elements a and b can be written in the form $sa + tb$ for some elements $s, t \in R$ (see Question 6, Exercises 6).

Definition 3.6. Elements a and b of an integral domain R are said to be *relatively prime* if the units of R are the only elements of R which divide both a and b.

In these circumstances 1 is an HCF of a and b and we can therefore write $(a, b) = 1$ to mean that a and b are relatively prime.

Corollary 3.7. *Let n be a positive integer and let $a \in \mathbb{Z}$. Then there exists $b \in \mathbb{Z}$ such that $ab \equiv 1 \pmod{n}$ if and only if $(a, n) = 1$.*

Proof. Suppose that $(a, n) = 1$. Then by 3.5 there exist $b, c \in \mathbb{Z}$ such that $ab + nc = 1$. Hence $ab \equiv 1 \pmod{n}$.

Conversely, suppose that $ab \equiv 1 \pmod{n}$. Then $ab + nc = 1$ for some $c \in \mathbb{Z}$. Now $(a, n) | a$ and $(a, n) | n$. So $(a, n) | 1$. Thus (a, n) is a unit of $\mathbb{Z}$ and we may take $(a, n) = 1$. ∎

Lemma 3.8. *Let R be a Euclidean ring and let a, b, c be non-zero elements of R such that $(a, b) = 1$. Suppose that $a | bc$. Then $a | c$.*

Proof. By 3.5 we have $1 = sa + tb$ for some $s, t \in R$. Hence $c = sac + tbc$. But a divides both sac and tbc. Therefore $a | c$. ∎

Notation 3.9. Let R be an integral domain and $a, b, n \in R$. We shall write $a \equiv b \pmod{n}$ if $n | a - b$. We have already used this congruence notation for $\mathbb{Z}$.

Theorem 3.10 (The Chinese remainder theorem). *Let R be a Euclidean domain. Let p, q be non-zero elements of R with $(p, q) = 1$. Let $a, b \in R$. Then there exists $x \in R$ such that $x \equiv a \pmod{p}$ and $x \equiv b \pmod{q}$.*

Proof. By 3.5 there exist $r, s \in R$ such that $1 = pr + qs$. Hence $a - b = pr(a - b) + qs(a - b)$. Therefore $a - pr(a - b) = b + qs(a - b) = x$, say. Thus $x \equiv a \pmod{p}$ and $x \equiv b \pmod{q}$. ∎

For the rest of this chapter we shall concentrate on the ring $\mathbb{Z}$ of integers.

Definition 3.11. A number $p \in \mathbb{Z}$ is called a *prime number* if $p \geq 2$ and p has no factors except ± 1 and $\pm p$.

Comments 3.12

(a) The first few prime numbers are $2, 3, 5, 7, 11, \ldots$

(b) Note that 1 is not a prime number.

(c) We shall study prime elements in other integral domains in Chapter 6.

We shall prove in 3.14 that every integer greater than 1 is a unique product of prime numbers.

Lemma 3.13. *Let p be a prime number and $b, c \in \mathbb{Z}$. Suppose that $p \mid bc$. Then $p \mid b$ or $p \mid c$.*

Proof. Suppose that p does not divide b. Then $(p, b) = 1$ since the only factors of p are ± 1 and $\pm p$. Hence $p \mid c$ by 3.8. ∎

It is easily seen that 3.13 is not true in general if p is not prime.

Theorem 3.14 (The fundamental theorem of arithmetic). *Every integer ≥ 2 is a product of primes which is unique apart from the order of the prime factors, where the phrase 'product of primes' includes a single prime.*

Proof. (a) *Existence:* Suppose that there exists an integer greater than 1, which is not a product of primes. We shall obtain a contradiction. Let n be the smallest integer such that $n \geq 2$ and n is not a product of primes. Then, in particular, n is not a prime. Hence $n = ab$ for some $a, b \in \mathbb{Z}$ with $2 \leq a < n$ and $2 \leq b < n$. Since a and b are smaller than n, each is a product of primes. Hence so is n. This is a contradiction. Therefore every integer ≥ 2 is a product of primes.

(b) *Uniqueness:* Let $n = p_1 p_2 \cdots p_r = q_1 q_2 \cdots q_s$ where p_i and q_j are primes (not necessarily distinct). We prove by induction on r that $r = s$ and after possibly renumbering the q_j we have $p_i = q_i$ for all i.

Suppose first that $r = 1$. Then $p_1 = q_1 q_2 \cdots q_s$. Since p_1 is prime we have $s = 1$ and $p_1 = q_1$.

Now assume that $r \geq 2$. We have $n = p_1 p_2 \cdots p_r = q_1 q_2 \cdots q_s$. So $p_1 \mid q_1 q_2 \cdots q_s$. Hence $p_1 \mid q_i$ for some i by repeated use of 3.13. Since q_i is a prime and $p_1 \neq 1$, we have $p_1 = q_i$. By renumbering the q_j if necessary, we may take $p_1 = q_1$. Therefore $p_2 p_3 \cdots p_r = q_2 q_3 \cdots q_s$. The left-hand side is a product of $r - 1$ primes. So by the induction hypothesis, $r - 1 = s - 1$. Thus $r = s$ and $p_2, \ldots, p_r$ are the same as $q_2, \ldots, q_r$ in some order. ∎

Comments 3.15

(a) Although one tends to take the unique factorization of integers into primes for granted, without it we would not be able to prove many standard results of number theory.

(b) In Chapter 6 we shall come across number systems where the unique factorization property does not hold.

(c) Theorem 3.14 is only a special case of Theorem 6.12 which proves the unique factorization property in a Euclidean domain.

Theorem 3.16 (Fermat's little theorem).   *Let p be a prime number and let a be an integer not divisible by p. Then*

$$a^{p-1} \equiv 1 \pmod{p}.$$

Proof.   Let S be the set of integers $\{a, 2a, \ldots, (p-1)a\}$. Let $s \in S$. Then $s = ia$ for some $i \in \mathbb{Z}$ with $1 \le i \le p-1$. Since p does not divide either i or a, it follows by 3.13 that p does not divide s. Thus no element of S is divisible by p. A similar argument shows that the elements of S belong to distinct congruence classes (mod p). Therefore they are congruent in some order to $1, 2, \ldots, (p-1)$ (mod p). However, the product of the elements of S is $(p-1)! \, a^{p-1}$. Therefore $(p-1)! \, a^{p-1} \equiv (p-1)! \pmod{p}$. Thus $p \,|\, (p-1)! \, (a^{p-1} - 1)$. Now p does not divide $(p-1)!$ since p is prime and each factor in $(p-1)!$ is less than p. Therefore $p \,|\, (a^{p-1} - 1)$ by 3.13. Thus $a^{p-1} \equiv 1 \pmod{p}$. ∎

For further material on number theory, the interested reader may consult an introductory text such as Burn (1982), Burton (1980), or Dudley (1978).

Exercises 3

1. Let a, b be non-zero elements of an integral domain R. Show that $a \,|\, b$ and $b \,|\, a$ if and only if $a = ub$ for some unit $u \in R$.

2. Let R be an integral domain. Let x, y be non-zero elements of R. Suppose that a and b are HCFs of x and y. Show that $a = ub$ for some unit $u \in R$.

3. Let $p, n_1, n_2, \ldots, n_k \in \mathbb{Z}$ where p is a prime number. Suppose that $p \,|\, n_1 n_2, \ldots, n_k$. Show that $p \,|\, n_i$ for some i.

4. Let p be a prime and $a \in \mathbb{Z}$. Show that $a^p \equiv a \pmod{p}$.

5. Find the remainder when 10^{500} is divided by 31.
     (*Hint*: Use Fermat's little Theorem.)

6. Let p be an odd prime and suppose that there are integers a and b, neither of which is divisible by p, such that p divides $a^2 + b^2$. Prove that $p \equiv 1 \pmod{4}$.

7. Find $x \in \mathbb{Z}$ with $0 \leq x \leq 27$ such that $17^{500} \equiv x$ (mod 28).

   (*Hint*: Use the Chinese remainder theorem.)

8. Prove that there are infinitely many primes which are congruent to 3 (mod 4).

   (*Hint*: Let k be the product of a finite number of such primes. Prove that there is at least one more such prime among the factors of $4k - 1$.)

9. Prove that there are infinitely many primes which are congruent to 1 (mod 4).

   (*Hint*: Let a be the product of a finite number of such primes. Take $b = 2$. Use question 6 to show that there is a prime $p \equiv 1$ (mod 4) such that p divides $a^2 + b^2$ but p does not divide a.)

   (Reading exercise) The following result was proved by Dirichlet (1805–59) and is known as Dirichlet's theorem for primes in arithmetic progressions.

   Let a and b be relatively prime positive integers. Then there are infinitely many primes among the terms of the arithmetic progression with first term a and common difference b, i.e. there are infinitely many primes of the form $a + nb$, where n is some positive integer. Questions 8 and 9 are special cases of this theorem (with $a = 3$ or 1 and $b = 4$). The proof of Dirichlet's theorem is hard. An easier proof for the special case when $a = 1$ is given in Niven and Powell (1976).

4 The four-squares theorem

We are now in a position to use the ring of Gaussian integers to prove one of the most famous theorems in number theory: every positive integer is the sum of four squares. The material of this chapter is a bit tricky and is not used later, so that the reader may omit it or read it superficially without worrying about the details. Nevertheless, we recommend that you look at Comments 4.7 and the Exercises.

Let us set the scene by considering some typical problems of classical number theory. Because for the time being we are in number-theory mode, when we say 'square' we mean 'square of non-negative integer' and similarly for cubes, etc. There has been a great deal of interest in questions about whether every positive integer can be represented as the sum of such-and-such a number of special integers; or which integers can be represented in this way; or can every large enough integer be represented in this way. Here are some examples.

Question: Is every even positive integer (except 2) the sum of two prime numbers?

Answer: Not known; the corresponding positive assertion is known as the Goldbach conjecture.

Question: Is every positive integer the sum of two squares?

Answer: No.

Question: Which positive integers are the sum of two squares?

Answer: See Exercises 6, Question 10.

Question: Is every positive integer the sum of three squares?

Answer: No; See Comments 4.7.

Question: Is every positive integer the sum of four squares?

Answer: Yes; we are about to prove this.

Question: Is every positive integer the sum of eight cubes?

Answer: No; but every integer greater than 239 is the sum of eight cubes (see Comments 4.7).

This list of questions, which could be greatly extended, should give the idea of the sort of thing which we have in mind. In this context we can appreciate the beautiful simplicity of the statement that every positive integer is the sum of four squares. This is one of the many influential assertions in number theory made by Fermat, but we do not know his proof (if indeed he had one). The earliest known proof is due to Lagrange.

Theorem 4.1 (Proved by Lagrange in 1770). *Every positive integer is the sum of four squares. Thus if n is a positive integer, there are integers a, b, c, d, such that $n = a^2 + b^2 + c^2 + d^2$.*

Let n be a positive integer. In order to prove that n is the sum of four squares, we may without loss of generality assume that n has no repeated prime factors. For example, if $n = 60$ we take out the largest square factor of n (namely 4), leaving 15, which has no repeated prime factors. Write 15 as the sum of four squares $15 = 9 + 4 + 1 + 1$ and now multiply through by the largest square factor of n to get $60 = 36 + 16 + 4 + 4$. From now on we assume that n is a positive integer with no repeated factors (see Question 8, Exercises 4).

Notation 4.2. If z is a complex number, $z = a + ib$ with $a, b \in \mathbb{R}$, we shall denote the complex conjugate of z by $\bar{z}$. Thus $\bar{z} = a - ib$. If B is the matrix

$$\begin{bmatrix} z_1 & z_2 \\ z_3 & z_4 \end{bmatrix}$$

then B^* will denote its conjugate transpose

$$\begin{bmatrix} \bar{z}_1 & \bar{z}_3 \\ \bar{z}_2 & \bar{z}_4 \end{bmatrix}.$$

The determinant of a matrix A will be denoted by $\det A$.

Strategy: Let p be a prime. We shall show that there are integers a and b such that $-1 \equiv a^2 + b^2 \pmod{p}$. We shall then use the Chinese remainder theorem to show that there are integers c and d such that $-1 \equiv c^2 + d^2 \pmod{n}$. Thus $1 + c^2 + d^2 = kn$ for some integer k. We shall then use the equation $1 + c^2 + d^2 = kn$ to construct an invertible 2×2 matrix A over the Gaussian integers, which has n in the $(1, 1)$-position. The final step is to write A in the form BB^* where B is a 2×2 matrix over the Gaussian integers and B^* is the complex conjugate of the transpose of B as above. Comparing the $(1, 1)$-entries of A and BB^* gives n as the sum of four squares.

Lemma 4.3. *Let p be a prime and let F denote the field of integers* (mod p). *Then every element of F is expressible in the form $x^2 + y^2$ for some $x, y \in F$. (For example,* (mod 7) *we can write* $0 = 0^2 + 0^2$, $1 = 1^2 + 0^2$, $2 = 1^2 + 1^2$, $3 = 3^2 + 1^2$, $4 = 2^2 + 0^2$, $5 = 2^2 + 1^2$ *and* $6 = 3^2 + 2^2$.)

Proof. The case $p = 2$ is trivial, so we suppose that p is odd. Set $S = \{x^2 : x \in F\}$. (For example, if $p = 7$ we have $S = \{0^2, 1^2, \ldots, 6^2\} = \{0, 1, 4, 2\}$.) Let F^* and S^* denote the sets of non-zero elements of F and S respectively. Define $f : F^* \to S^*$ by $f(x) = x^2$ for all $x \in F^*$. Clearly $f(xy) = f(x)f(y)$ for all $x, y \in F^*$. Thus f is a surjective group homomorphism from the multiplicative group F^* to the multiplicative group S^*. We have $f(x) = 1 \Leftrightarrow x^2 = 1 \Leftrightarrow (x - 1)(x + 1) = 0$. Therefore the kernel of f, $\mathrm{Ker}(f) = \{x \in F^* | f(x) = 1\} = \{1, -1\}$. Thus F^* has $p - 1$ elements and $\mathrm{Ker}(f)$ has 2 elements. However, by the homomorphism theorem of groups, $S^* = f(F^*)$ is isomorphic to $F^*/\mathrm{Ker}(f)$. Hence S^* and $F^*/\mathrm{Ker}(f)$ have the same number of elements. Therefore, by Lagrange's theorem on groups, S^* has $(p - 1)/2$ elements, so S has $(p - 1)/2 + 1 = (p + 1)/2$ elements. Let z be a fixed element of F and set $A = \{z - s | s \in S\}$. The subsets A and S of F each have $(p + 1)/2$ elements and F has p elements. Hence the intersection of A and S is non-empty. Therefore there are elements s and t of S such that $z - s = t$. However, $s = x^2$ and $t = y^2$ for some $x, y \in F$. Therefore $z = x^2 + y^2$, as required. ∎

Corollary 4.4. *There are integers a and b such that $-1 \equiv a^2 + b^2$ (mod p).*

Proof. In the proof of 4.3 take $z = -1$ and let a and b be integers which belong to the congruence classes x and y. ∎

Lemma 4.5. *Given n, a positive integer with no repeated prime factors, there are integers c, d, k such that $1 + c^2 + d^2 = kn$.*

Proof. The case $n = 1$ is trivial. So assume $n > 1$. We can write $n = p_1 p_2 \cdots p_r$ where each p_i is prime and $p_i \neq p_j$ if $i \neq j$. We shall proceed by induction on the positive integer r. If $r = 1$ then n is prime and the result follows from 4.4. Suppose now that $r \neq 1$. Set $p = p_1$ and $q = p_2 p_3 \cdots p_r$. By 4.4 there are integers a, b, and u such that $1 + a^2 + b^2 = up$. Since q is a product of $r - 1$ distinct primes, the induction hypothesis gives that there are integers e, f and v such that $1 + e^2 + f^2 = vq$. However, p and q are relatively prime. Therefore by the Chinese remainder theorem (3.10), there exists $c \in \mathbb{Z}$ such that $c \equiv a$ (mod p) and $c \equiv e$ (mod q). Similarly there is an integer d such that $d \equiv b$ (mod p) and $d \equiv f$ (mod q). Therefore (mod p) we have

$$1 + c^2 + d^2 \equiv 1 + a^2 + b^2 \equiv up \equiv 0.$$

Similarly (mod q) we have

$$1 + c^2 + d^2 \equiv 1 + e^2 + f^2 \equiv vq \equiv 0.$$

Thus $1 + c^2 + d^2$ is divisible by both p and q. However, p and q are relatively prime and $pq = n$. Therefore n divides $1 + c^2 + d^2$. Hence $1 + c^2 + d^2 = kn$ for some integer k. ∎

Keeping to the notation of 4.5 and noting that k must be positive, we write

$$A = \begin{bmatrix} n & c + \mathrm{i}d \\ c - \mathrm{i}d & k \end{bmatrix}.$$

Then A has the following properties: (i) The entries of A are Gaussian integers, (ii) det $A = 1$, (iii) The diagonal entries of A are positive integers, and (iv) $A = A^*$.

We shall show in 4.6 that there exists a matrix

$$B = \begin{bmatrix} q & r \\ s & t \end{bmatrix}$$

such that $A = BB^*$ and q, r, s, t are Gaussian integers. Comparing the (1,1)-entries of A and BB^* gives $n = q\bar{q} + r\bar{r}$. However, there are integers u, v, w, and x such that $q = u + \mathrm{i}v$ and $r = w + \mathrm{i}x$. Therefore

$$n = q\bar{q} + r\bar{r} = u^2 + v^2 + w^2 + x^2.$$

Thus the proof of the four-squares theorem will be complete when we have proved 4.6.

Lemma 4.6.   *Let J denote the ring of Gaussian integers and let*

$$A = \begin{bmatrix} u & x \\ \bar{x} & v \end{bmatrix},$$

where u and v are positive integers and $x \in J$. Suppose further that det $A = 1$. Then there is a 2×2 matrix B with entries in J such that $A = BB^*$.

Proof.   We associate a non-negative integer $n(A)$ with A. Define $n(A) = x\bar{x}$. We shall proceed by induction on $n(A)$. Suppose firstly that $n(A) = 0$. Then $x = 0$ and the fact that det $A = 1$ gives $uv = 1$. Hence $u = v = 1$ and we can take B to be the 2×2 identity matrix.

Suppose now that $n(A)$ is positive. We consider four cases:

Case 1: $u = 1$. Since $\det A = 1$, we have $v = 1 + x\bar{x}$ and we can take

$$B = \begin{bmatrix} 0 & 1 \\ 1 & \bar{x} \end{bmatrix}.$$

Case 2: $v = 1$. Similar to Case 1.

Case 3: $2 \le u \le v$. Set

$$C = \begin{bmatrix} 1 & 0 \\ y & 1 \end{bmatrix},$$

where y is for the moment an arbitrary element of $\mathbf{J}$. Set $X = CAC^*$. We shall first show that X has the same four properties as A, i.e. properties (i),…,(iv) described above. Clearly the entries of X belong to $\mathbf{J}$. Also $\det X = \det C \cdot \det A \cdot \det C^* = 1 \cdot 1 \cdot 1 = 1$. Let z denote the $(1,2)$-entry of X. Then $z = u\bar{y} + x$. The $(2,1)$-entry of X is $uy + \bar{x} = \overline{(u\bar{y} + x)} = \bar{z}$. Let u' and v' denote the diagonal entries of X. Then $u' = u$, which is a positive integer; $v' = uy\bar{y} + \overline{xy} + xy + v$ is a real Gaussian integer and so is an integer. Finally the fact that $u'v' - z\bar{z} = \det X = 1$ shows that v' is positive. Thus X has the same properties as A.

We next show that y can be chosen so that $n(X) < n(A)$, i.e. $z\bar{z} < x\bar{x}$. As in 2.10 let $N(a) = a\bar{a}$ for $a \in \mathbf{J}$. Thus we want to choose y so that $N(z) < N(x)$, i.e. $N(u\bar{y} + x) < N(x)$, i.e. $N(x/u + \bar{y}) < N(x/u)$. We can arrange this as follows: Since $2 \le u \le v$ and $1 = \det A = uv - x\bar{x}$ we have $N(x/u) = (x/u)\overline{(x/u)} = x\bar{x}/u^2 = (uv - 1)/u^2 \ge (u^2 - 1)/u^2 \ge 1 - \frac{1}{4} = \frac{3}{4}$. By 2.10 there is a Gaussian integer q such that $N(x/u - q) \le \frac{1}{2}$. Therefore $N(x/u - q) < N(x/u)$.

Set $y = -\bar{q}$. With this choice of y we have $n(X) < n(A)$. By induction we have $X = DD^*$ for some 2×2 matrix D with entries in $\mathbf{J}$. Set $E = C^{-1}$. Then $E^* = (C^*)^{-1}$ and $A = EDD^*E^* = ED(ED)^*$. Now take $B = ED$.

Case 4: $2 \le v \le u$. This is similar to Case 3 taking

$$C = \begin{bmatrix} 1 & y \\ 0 & 1 \end{bmatrix}.$$

We illustrate Case 3 by means of an example. Take

$$A = \begin{bmatrix} 2 & 3 + 4i \\ 3 - 4i & 13 \end{bmatrix}.$$

Then $u = 2$, $v = 13$ and $x = 3 + 4i$. We have $N(A) = x\bar{x} = 25$. To find y we let q be a Gaussian integer nearest to $x/u = \frac{3}{2} + 2i$ and we take $q = 1 + 2i$.

(We could also take $q = 2 + 2i$.) We set $y = -\bar{q} = -1 + 2i$. This gives

$$X = \begin{bmatrix} 2 & 1 \\ 1 & 1 \end{bmatrix}$$

and $n(X) = 1$. As in Case 2 we write $X = DD^*$ with

$$D = \begin{bmatrix} 1 & 1 \\ 1 & 0 \end{bmatrix}.$$

We have

$$E = C^{-1} = \begin{bmatrix} 1 & 0 \\ 1 - 2i & 1 \end{bmatrix}$$

and $A = BB^*$ with

$$B = ED = \begin{bmatrix} 1 & 1 \\ 2 - 2i & 1 - 2i \end{bmatrix}.$$

Comparing the diagonal entries of A and BB^* gives $2 = 1 + 1$ and $13 = (2 - 2i)(\overline{2 - 2i}) + (1 - 2i)(\overline{1 - 2i}) = 2^2 + 2^2 + 1^2 + 2^2$.   ∎

Comments 4.7. It is worth noting here that the proof of the four-squares theorem did not require the unique factorization property for the Gaussian integers.

There are many proofs of the four-squares theorem. Three more are given in Hardy and Wright (1979). Now that we know that every positive integer n can be written as a sum of four squares, we can ask about how many different ways there are of doing it (and what do we mean by 'different ways'?). For example $26 = 25 + 1 + 0 + 0 = 16 + 9 + 1 + 0 = 9 + 9 + 4 + 4$. Hardy and Wright (1979) contains more information on this.

Not every positive integer is the sum of three squares. The squares less than 7 are 0, 1 and 4. It is not possible to write 7 as the sum of three of them. The three squares theorem states that n is a sum of three squares if and only if n is not of the form $4^t(8k + 7)$ for non-negative integers t and k. The 'only if' part is a quite easy exercise but the 'if' part is very hard. It is rather easy to characterize those n which can be written as a sum of two squares. This is done in 6.15 and Question 10, Exercises 6.

There are many results and problems in number theory concerned with sums of powers higher than squares. For example, it is known that every positive integer is the sum of nine cubes. It is easy to check that 23 and 239 require nine cubes (i.e. eight will not do) and it is known that 23 and 239 are the only positive integers which require nine cubes. It is conjectured, but not proved, that every positive integer is the sum of 19 fourth powers.

It is known that every positive integer is the sum of 37 fifth powers, and there are various other conjectures and results. This material in number theory usually comes under the heading 'Waring's problem'. The interested reader might like to consult Hardy and Wright (1979), Small (1977) and references given there.

Exercises 4

1. Show that the square of every integer is congruent to 0 or 1 (mod 4).

2. Show that the square of an even integer is congruent to 0 or 4 (mod 8) and that the square of an odd integer is congruent to 1 (mod 8).

3. Use Question 2 to show that any number which is congruent to 7 (mod 8) is not the sum of three squares.

4. Let $n = 4^t(8k + 7)$ where t and k are non-negative integers. Show that n is not the sum of three squares.

   (*Hint*: Use induction on t. Show that if $n = a^2 + b^2 + c^2$ then a, b, c are all even if t is positive).

5. Verify that the numbers 23 and 239 can be written as the sum of 9 but not 8 cubes.

6. Verify that 79 can be written as the sum of 19 but not 18 fourth powers.

7. Verify that 223 can be written as the sum of 37 but not 36 fifth powers.

8. Show that if every positive integer which has no repeated prime factor is a sum of four squares, then the same is true of any positive integer.

5 Fields and polynomials

The two most important results which we shall prove in this chapter are the division algorithm for polynomials in one indeterminate over a field, and that the integers modulo p form a field for every prime number p (thus providing our first examples of finite fields).

We gave the definition of a field earlier in Chapter 2. Fields and polynomials over them play an important role in algebra. In this chapter we collect together some of the elementary but important properties which we shall need later.

Recall from 2.4 that a field is a ring R such that (i) $1 \neq 0$ in R and (ii) every non-zero element of R is a unit of R. Thus a field is a ring in which the non-zero elements form a group under multiplication.

Examples 5.1. The following are easily seen to be fields:

(a) the rational numbers $\mathbb{Q}$;

(b) the real numbers $\mathbb{R}$;

(c) the complex numbers $\mathbb{C}$. ∎

Non-examples 5.2. The following are not fields:

(a) $\mathbb{Z}$ the ring of integers;

(b) $\mathbb{Z}[X]$ the ring of polynomials over $\mathbb{Z}$;

(c) the ring $\mathbb{Z}/6\mathbb{Z}$. The elements 2 and 3 are non-zero elements of this ring. However $2 \cdot 3 = 0$ in $\mathbb{Z}/6\mathbb{Z}$. So neither 2 nor 3 can be a unit of $\mathbb{Z}/6\mathbb{Z}$. ∎

Comments 5.3

(a) We shall show in 5.7 that $\mathbb{Z}/n\mathbb{Z}$ is a field if and only if n is prime.

(b) Every field is an integral domain (Question 1, Exercises 5).

(c) Not every integral domain is a field, for example $\mathbb{Z}$.

(d) The definitions 'ring', 'integral domain', 'field' are successively more restrictive.

Proposition 5.4.   *Let a and n be integers with n positive. Then a is a unit of the ring $\mathbb{Z}/n\mathbb{Z}$ if and only if a and n are relatively prime.*

Proof.   Firstly suppose that $(a, n) = 1$. Then $sa + tn = 1$ for some integers s and t by 3.5. In $\mathbb{Z}/n\mathbb{Z}$ this gives $sa = 1$. Thus a is a unit with inverse s in $\mathbb{Z}/n\mathbb{Z}$. Conversely, suppose that a is a unit of $\mathbb{Z}/n\mathbb{Z}$. Then there is an integer b such that in $\mathbb{Z}/n\mathbb{Z}$ we have $ab = 1$. Thus in $\mathbb{Z}$ we have $ab = 1 + kn$ for some integer k. Since $1 = ab - kn$ it follows that a and n have no non-trivial common factors.   ∎

Examples 5.5

(a) The elements of $\mathbb{Z}/2\mathbb{Z}$ are 0 and 1 and of these the only unit is 1.

(b) The elements of $\mathbb{Z}/3\mathbb{Z}$ are 0, 1 and 2 and of these the units are 1 and 2. The element 2 is its own inverse because in $\mathbb{Z}/3\mathbb{Z}$ we have $2 \cdot 2 = 4 = 1$.

(c) The elements of $\mathbb{Z}/4\mathbb{Z}$ are 0, 1, 2 and 3 and of these the units are 1 and 3.

(d) The elements of $\mathbb{Z}/5\mathbb{Z}$ are 0, 1, 2, 3 and 4 and of these the units are 1, 2, 3 and 4. In $\mathbb{Z}/5\mathbb{Z}$ we have $2 \cdot 3 = 6 = 1$, so that 2 and 3 are inverses for each other. The units of $\mathbb{Z}/5\mathbb{Z}$ form a group of order 4 (i.e. a group with 4 elements). This group is cyclic because all the units of $\mathbb{Z}/5\mathbb{Z}$ can be expressed as powers of the single element 2, namely $1 = 2^0$, $2 = 2^1$, $3 = 8 = 2^3$, $4 = 2^2$.

(e) The units of $\mathbb{Z}/6\mathbb{Z}$ are 1 and 5, i.e. 1 and -1.

(f) All the non-zero elements of $\mathbb{Z}/7\mathbb{Z}$ are units. Again the group of units is cyclic. This time 2 is not a generator but 3 is. For in $\mathbb{Z}/7\mathbb{Z}$ we have $2^3 = 1$, so that the only distinct powers of 2 are 1, 2 and 4; but the distinct powers of 3 are 1, 3, $3^2 = 9 = 2$, $3^3 = 6$, $3^4 = 18 = 4$ and $3^5 = 12 = 5$.   ∎

We shall show later (10.8) that the group of units of $\mathbb{Z}/n\mathbb{Z}$ is cyclic whenever n is prime.

Example 5.6.   Let U be the group of units of $\mathbb{Z}/12\mathbb{Z}$. We denote the elements of $\mathbb{Z}/12\mathbb{Z}$ by $0, 1, 2, \ldots, 11$. Of these the only ones which are relatively prime to 12 are 1, 5, 7, 11. Thus $U = \{1, 5, 7, 11\} = \{\pm 1, \pm 5\}$. Thus U has order 4. In $\mathbb{Z}/12\mathbb{Z}$ we have $5^2 = 25 = 1$. It follows that $u^2 = 1$ for all $u \in U$. Thus U is the Klein 4-group. Note particularly that U is not cyclic.   ∎

Theorem 5.7. *Let* $n \geq 2$ *be a positive integer. Then the following are equivalent*:

(i) n *is a prime number.*

(ii) $\mathbb{Z}/n\mathbb{Z}$ *is an integral domain.*

(iii) $\mathbb{Z}/n\mathbb{Z}$ *is a field.*

Proof. (ii) $\Rightarrow$ (i) Suppose that n is not prime. Then $n = ab$ for some positive integers a, b less than n. In $\mathbb{Z}/n\mathbb{Z}$ we have $a \neq 0$ and $b \neq 0$ but $ab = n = 0$. Thus $\mathbb{Z}/n\mathbb{Z}$ is not an integral domain. Hence if $\mathbb{Z}/n\mathbb{Z}$ is an integral domain then n is a prime.

(i) $\Rightarrow$ (iii) Let a be an integer which is non-zero in $\mathbb{Z}/n\mathbb{Z}$. Then a is not divisible by n. Since n is prime we then have $(a, n) = 1$. Therefore by 5.4 a is a unit of $\mathbb{Z}/n\mathbb{Z}$.

(iii) $\Rightarrow$ (ii) A field is automatically an integral domain (see Question 1, Exercises 5). ∎

At present our examples of fields are $\mathbb{Q}$, $\mathbb{R}$, $\mathbb{C}$ and $\mathbb{Z}/p\mathbb{Z}$ with p prime. Of these only $\mathbb{Z}/p\mathbb{Z}$ is finite. We shall construct further finite fields later in Chapters 14 and 15. Also we shall show in Chapter 7 that any integral domain is contained in a field in the same sort of way that $\mathbb{Z}$ is contained in $\mathbb{Q}$.

Next we study the polynomial ring $F[X]$ where F is a field. The results we shall prove for polynomials in one indeterminate over a field may not be true for such polynomial rings as $\mathbb{Z}[X]$ and $\mathbb{Q}[X, Y]$. First we restate two theorems proved earlier. Recall that $\partial(f(X))$ denotes the degree of the polynomial $f(X)$.

Theorem 5.8 (The division algorithm). *Let F be a field and let $f(X)$, $g(X) \in F[X]$ with $g(X) \neq 0$. Then there exist $q(X)$ and $r(X) \in F[X]$ such that*

$$f(X) = q(X)g(X) + r(X),$$

where either $r(X) = 0$ or $\partial(r(X)) < \partial(g(X))$.

Proof. See 2.9. ∎

Corollary 5.9. $F[X]$ *is Euclidean.*

Proof. See 2.9. ∎

Note that the units of $F[X]$ are the non-zero constants, i.e. the non-zero elements of F. This should be contrasted with the fact that the only units of $\mathbb{Z}[X]$ are 1 and -1. Note also that $\mathbb{Z}[X]$ is not Euclidean (Question 7, Exercises 2).

Proposition 5.10. *Let F be a field, $a \in F$ and $f(X) \in F[X]$. Then $f(a) = 0$ if and only if $f(X) = (X - a)g(X)$ for some $g(X) \in F[X]$.*

It is important to note that a and all the coefficients of $f(X)$ and $g(X)$ belong to the same field F.

Proof. If $f(X) = (X - a)g(X)$ then clearly $f(a) = 0$. Conversely suppose that $f(a) = 0$. By 5.8 there exist $g(X)$ and $k(X) \in F[X]$ such that $f(X) = (X - a)g(X) + k(X)$ where either $k(X) = 0$ or $\partial(k(X)) < \partial(X - a) = 1$. In either case $k(X)$ is a constant. Since $f(a) = 0$ we have $k(a) = 0$. Now as $k(X)$ is a constant it follows that $k(X) = 0$. Thus $f(X) = (X - a)g(X)$. ∎

Theorem 5.11. *Let $f(X)$ be a non-zero polynomial of degree n with coefficients in a field F. Then the equation $f(X) = 0$ has at most n distinct solutions in F.*

Proof. We shall prove this by induction on n. When $n = 1$ the polynomial $f(X)$ is linear and the equation $f(X) = 0$ has exactly one root. Now assume that $n \geq 2$. We have nothing to prove if $f(X)$ has no roots in F. Suppose now that $a \in F$ is a root of $f(X)$. Then by 5.10 we have $f(X) = (X - a)g(X)$ for some $g(X) \in F[X]$. Clearly $\partial(g(X)) = n - 1$. By the induction hypothesis $g(X)$ has at most $n - 1$ distinct roots in F. Since $F[X]$ is an integral domain any root of $f(X)$ other than a is also a root of $g(X)$. It follows that $f(X)$ has at most n distinct roots in F. ∎

Comments 5.12

(a) In Question 9, Exercises 1 there is an example of a quadratic equation which has 4 distinct roots in $\mathbb{Z}/6\mathbb{Z}$. Thus 5.11 is false if F is replaced by an arbitrary ring.

(b) Let R be an arbitrary integral domain. We shall show in Chapter 7 that R is contained in a field F. It will then follow from 5.11 that if $f(X)$ is a non-zero polynomial with coefficients in the integral domain R then the equation $f(X) = 0$ has at most $\partial(f(X))$ distinct roots in R.

Exercises 5

1. Prove that every field is an integral domain.

2. Let G be the group of units of $\mathbb{Z}/10\mathbb{Z}$. List the elements of G. Prove that G is cyclic.

3. Let G be the group of units of $\mathbb{Z}/20\mathbb{Z}$. List the elements of G, pairing each element with its inverse. Determine whether or not G is cyclic.

4. Let R be a finite integral domain. Prove that R is a field.
   (*Hint*: Fix a non-zero element a of R and consider the set of all elements ab where b runs over R).

5. Working in $\mathbb{Q}[X]$ let

$$a(X) = 10X^5 - 4X^2 + 3 \quad \text{and} \quad b(X) = 6X^2 + 2X + 5.$$

Find $q(X)$ and $r(X) \in \mathbb{Q}[X]$ such that

$$a(X) = q(X)b(X) + r(X),$$

where either $r(X) = 0$ or $\partial(r) < \partial(b)$.

6. (Wilson's theorem) Let p be a prime number. Show that $(p-1)! \equiv -1 \pmod{p}$.
   (*Hint*: Let $F = \mathbb{Z}/p\mathbb{Z}$. Working over F set

$$f(X) = X^{p-1} - 1 - (X-1)(X-2)(X-3) \cdots (X-(p-1)).$$

Use Fermat's little theorem to show that $f(a) = 0$ for each of the $p-1$ non-zero elements $a \in F$. Use 5.11 to show that all the coefficients of $f(X)$ are 0. Deduce the result by considering the constant term of $f(X)$).

6 Unique factorization domains

At first it is perhaps rather difficult to appreciate why so much importance is placed on the result that every positive integer (except 1) is a unique product of prime numbers (see 3.14). It is known as the fundamental theorem of arithmetic, which suggests that it must be very significant, but in practice we take it for granted that, for instance, the only way of writing 10 as a product of prime numbers is as $2 \cdot 5$ (or $5 \cdot 2$, but that is really the same thing). We might hope that this sort of unique factorization property, when suitably defined, also holds in such systems as the Gaussian integers. If it does, we will have to prove it; we have no right to take it for granted. In fact we shall show that unique factorization does hold for Gaussian integers, but we shall also give examples (one of them apparently rather similar to the Gaussian integers) where unique factorization fails. It is these examples of failure which may bring home to us for the first time the need to be careful. Unique factorization should not be taken for granted, and when it does hold we should appreciate its value in underpinning some of the arithmetic arguments which we may want to use.

Historically, the unique factorization property (or its failure) has been of great interest in number theory. To pick just one example and to over-simplify a very complicated story: one of the reasons why Fermat's last theorem was so hard to prove was that the number systems that are normally used to tackle it do not all have unique factorization (for more details see Edwards (1977)). Once we have accepted that this highly desirable property can fail, it makes it worth while to put in the effort to establish it when it does hold (and this is often very difficult to do). In this spirit we shall show that every Euclidean ring (in particular, the Gaussian integers) has the unique factorization property. This will enable us to prove another of Fermat's assertions, namely that every prime number of the form $4n + 1$ is the sum of two squares.

In this chapter we shall study domains in which each non-trivial element can be written uniquely as a product of elements which correspond to prime numbers. For these purposes, a non-trivial element is one which is not 0 and which is not a unit. Concerning uniqueness we want to ensure, for example, that the following are counted as being the same factorization of 6 in $\mathbb{Z}$: $6 = 2 \cdot 3 = 3 \cdot 2 = -2 \cdot -3 = -3 \cdot -2$. The element which corresponds to a prime number is one which does not factorize except in trivial ways (see 6.1). The technical word for this is 'irreducible' as defined below.

For general integral domains the term 'prime element' has a different meaning, which we shall come to later.

Definition 6.1. An element x of an integral domain R is said to be *irreducible* if (i) $x \neq 0$, (ii) x is not a unit of R, and (iii) x does not factorize properly in R. Thus if x is irreducible and $x = yz$ for some $y, z \in R$, then either y or z is a unit.

Example 6.2. The irreducible elements of $\mathbb{Z}$ are the prime numbers $2, 3, 5, 7, 11, \ldots$ and their negatives $-2, -3, -5, \ldots$.   ■

Example 6.3. It is a consequence of the fundamental theorem of algebra that every non-constant polynomial in $\mathbb{C}[X]$ is a product of linear factors. Hence the irreducible elements of $\mathbb{C}[X]$ are its linear elements. The corresponding statements are not true for polynomials over $\mathbb{Q}$ or $\mathbb{R}$ (see Question 8, Exercises 6, and Chapter 8).   ■

Definition 6.4. An integral domain R is said to be a *unique factorization domain (UFD)* if every non-zero, non-unit element of R is expressible uniquely as a product of irreducible elements of R (where we disregard changes in the order of the factors and also disregard changes made by multiplying factors by units).

Example 6.5. $\mathbb{Z}$ is a UFD since every positive integer except 1 is a unique product of prime numbers (3.14). From this it easily follows that every 'non-trivial' element of $\mathbb{Z}$ is a unique product of irreducible elements. For example $-6 = 2 \cdot -3$ where 2 and -3 are irreducible elements of $\mathbb{Z}$, and this is the only factorization of -6. By 'non-trivial' element we mean an element which is not 0 and is not a unit, so that in $\mathbb{Z}$ the trivial elements are 0 and the units 1 and -1.   ■

Example 6.6. Let $R = \mathbb{C}[X, Y]$. Thus R is the ring of all polynomials in two indeterminates X and Y with coefficients in $\mathbb{C}$. Let S be the subset of R consisting of those elements of R which are unaltered when X is changed to $-X$ and Y is changed to $-Y$. For example, S contains X^2, XY and Y^2 but does not contain X or Y. We leave it as an exercise to show that S is a subring of R. The non-zero elements of $\mathbb{C}$ belong to S and are the only units of S. Clearly the elements X^2, XY and Y^2 of S do not factorize properly in S. (Remember that X and Y are not in S). Thus X^2, XY and Y^2 are irreducible elements of S. Now, $X^2 \cdot Y^2$ and $XY \cdot XY$ are genuinely different factorizations of $X^2 Y^2$ into irreducible factors. Therefore S is not a UFD. We note that by 8.14(ii) R itself is a UFD.   ■

Example 6.7. Let R be the subring of $\mathbb{Z}[X]$ consisting of all polynomials of the form $a_0 + a_1 X + \cdots + a_n X^n$ where $a_i \in \mathbb{Z}$ for all i and a_i is even for all $i \neq 0$. The trivial elements of R are 0 together with the units 1 and -1. Clearly the only factorizations of $2X$ in $\mathbb{Z}[X]$ are $2 \cdot X$, $-2 \cdot X$, $X \cdot -2$ and $-X \cdot -2$. None of these are allowable in R because X and $-X$ are not in R. Therefore $2X$ is an irreducible element of R. Similarly $2X^2$ is an irreducible element of R, as also is 2. Hence $4X^2 = 2X \cdot 2X = 2 \cdot 2X^2$ gives two genuinely different factorizations of $4X^2$ into irreducible factors, because the factors are irreducible and it is impossible to turn $2X$ into either 2 or $2X^2$ by multiplying by a unit of R. ∎

Our next major aim is to show that every Euclidean ring is a UFD.

Lemma 6.8. *Let R be a Euclidean ring and let x be an irreducible element of R. Suppose that $a, b \in R$ and $x \mid ab$. Then $x \mid a$ or $x \mid b$.*

Proof. Suppose that x does not divide a. Let $d \in R$ such that $d \mid x$ and $d \mid a$. In particular, $x = dr$ for some $r \in R$. Since x is irreducible, either d or r is a unit of R. Suppose that r is a unit of R. We have $a = dw$ for some $w \in R$. Thus $a = dw = xr^{-1}w$ with $r^{-1}w \in R$.

Hence $x \mid a$ which is a contradiction. Therefore d is a unit of R. Thus x and a are relatively prime. Since $x \mid ab$ it follows by 3.8 that $x \mid b$. ∎

Definition 6.9. An element p of an integral domain R is said to be *prime* if (i) $p \neq 0$, (ii) p is not a unit of R and (iii) whenever a and b are elements of R such that $p \mid ab$ then $p \mid a$ or $p \mid b$.

Comments 6.10

(i) A prime element of an integral domain is always irreducible (see Question 2, Exercises 6).

(ii) An irreducible element of an integral domain need not be prime. For example the element XY in 6.6 is irreducible but is not prime because it divides the product of X^2 and Y^2 without dividing either X^2 or Y^2 separately.

(iii) In contrast to (ii) the significance of 6.8 is that every irreducible element of a Euclidean domain is prime.

Lemma 6.11. *Let R be a Euclidean ring with respect to the function N. Let a and b be non-zero elements of R such that b is not a unit of R. Then $N(ab) > N(a)$.*

Proof. We have $a = qab + r$ for some $q, r \in R$ where either $r = 0$ or $N(r) < N(ab)$. Suppose that $r = 0$. Then $a = qab$. Since $a \neq 0$, it follows that $qb = 1$. Thus b is a unit of R which is a contradiction. Therefore $r \neq 0$. Hence $N(r) < N(ab)$. However, $r = a(1 - qb)$. Since $r \neq 0$, we have $1 - qb \neq 0$. Therefore, $N(ab) > N(r) = N(a(1 - qb)) \geq N(a)$ by 2.6(ii).   ∎

Theorem 6.12. *Every Euclidean ring is a UFD.*

Proof. Let R be a Euclidean ring with respect to the function N. Let x be a non-zero, non-unit element of R. We shall first show that x is the product of irreducible elements of R, and then consider the uniqueness of the factors.

Suppose to the contrary that x is not a product of irreducible elements of R. Let S be the set of all non-zero, non-unit elements of R, which are not products of irreducible elements. Then S is non-empty since $x \in S$. Let $a \in S$ be such that $N(a)$ is as small as possible. We can do this since $N(r)$ is a non-negative integer for all $0 \neq r \in R$. Then, in particular, a itself is not irreducible (by convention, the phrase 'x is a product of irreducible elements' includes the case when there is just one irreducible factor, i.e., the case when x itself is irreducible). Therefore, $a = bc$ for some non-zero, non-units $b, c \in R$. By 6.11, $N(a) = N(bc) > N(b)$. Similarly, $N(a) > N(c)$. By the choice of a we have $b \notin S$ and $c \notin S$. Therefore each of b and c is a product of irreducible elements. However, $a = bc$. Hence a is a product of irreducible elements, which is a contradiction since $a \in S$. Thus the assumption that x is not a product of irreducible elements has led to a contradiction.

Now let x be again a non-zero non-unit of R. Suppose that $x = p_1 p_2 \cdots p_r = q_1 q_2 \cdots q_s$ where each p_i and q_j is irreducible. We must show that $r = s$ and that the p_i are the same as the q_j except possibly for their order and multiplication by units. Since $p_1 | x$ we know that $p_1 | q_1 q_2 \cdots q_s$. By 6.8 p_1 is prime and so it follows that $p_1 | q_i$ for some i. Renumbering the q_i if necessary, we may suppose that $p_1 | q_1$. Thus $q_1 = p_1 w$ for some $w \in R$. Now q_1 is irreducible and p_1 is not a unit. Therefore w is a unit. Since $p_1 p_2 \cdots p_r = q_1 q_2 \cdots q_s = w p_1 q_2 \cdots q_s$ we have $p_2 p_3 \cdots p_r = w q_2 q_3 \cdots q_s$. As w is a unit, we may without loss of generality rename $w q_2$ as q_2. Thus $p_2 p_3 \cdots p_r = q_2 q_3 \cdots q_s$. Hence $p_2 | q_2 q_3 \cdots q_s$. As above, without loss of generality we may assume that $q_2 = p_2$ and so on. Hence we obtain that $r \leq s$, and by symmetry $r = s$. Also for each i, we have $p_i = u_i q_i$, where u_i is a unit of R. Thus R is UFD.   ∎

Comments 6.13

(a) All rings which were shown in Chapter 2 and its exercises to be Euclidean are UFDs. We shall shortly use the fact that the Gaussian integers are a UFD to prove one of Fermat's results in number theory.

(b) Some readers may be familiar with the method for finding HCFs in $\mathbb{Z}$ using Euclid's algorithm. This method can be extended in a natural way to find HCFs in any Euclidean domain.

Proposition 6.14 (Euler's criterion). *Let p be an odd prime and let a be an integer which is not divisible by p. Then there is an integer x such that $a \equiv x^2 \pmod{p}$ if and only if $a^{(p-1)/2} \equiv 1 \pmod{p}$.*

Proof. We shall work in $F = \mathbb{Z}/p\mathbb{Z}$. Set $r = (p-1)/2$. Note that r is an integer because p is odd. We are given that $a \neq 0$ and we must show that $a = x^2$ for some $x \in F$ if and only if $a^r = 1$. Let S be the set of all non-zero elements $b \in F$ such that $b = x^2$ for some $x \in F$. Let U be the set of elements $b \in F$ such that $b^r = 1$. We shall show that $S = U$ and that will complete the proof.

Firstly let $s \in S$. Then $s = x^2$ for some $x \in F$. We have $x \neq 0$ because $s \neq 0$. Hence by Fermat's little theorem (3.16) we have $1 = x^{p-1} = x^{2r} = s^r$, so that $s \in U$. Thus $S \subseteq U$. The elements $1^2, 2^2, \ldots, r^2$ of F are clearly elements of S. We next show that they are distinct. Suppose that $i^2 = j^2$ in F for some integers i and j in the range from 1 to r. Then $0 = i^2 - j^2 = (i+j)(i-j)$. Since F is an integral domain it follows that $i + j = 0$ or $i - j = 0$. Now $i + j$ is in the range from 2 to $2r = p - 1$, so that $i + j \neq 0$ in F. Hence $i = j$. Therefore we know that S contains at least r distinct elements. On the other hand, by 5.11 U contains at most r elements because the elements of U are roots of a polynomial equation over the field F of degree r, namely $X^r = 1$. Since $S \subseteq U$ it follows that $S = U$. ■

The following theorem was stated by Fermat and is key to the two-squares theorem (see Question 10, Exercises 6). The earliest surviving proof is due to Euler.

Theorem 6.15 (Fermat). *Let p be a prime such that $p \equiv 1 \pmod{4}$. Then p is the sum of two squares.*

Proof. We have $p = 4n + 1$ for some positive integer n. Thus $(-1)^{(p-1)/2} = (-1)^{2n} = 1$. Therefore by Euler's criterion (6.14) there is an integer k such that $-1 \equiv k^2 \pmod{p}$. Thus $p \mid k^2 + 1$. We shall now work in the Gaussian integers so that we can factorize $k^2 + 1$ as $(k + i)(k - i)$.

Let J be the ring of Gaussian integers and let N be the usual norm function on J. We know that J is UFD and that the units of J are $1, -1, i, -i$. Also from the above we have $k^2 + 1 = wp$ for some $w \in \mathbb{Z}$. As $w \in J$ we know that $p \mid k^2 + 1$ in J. However, p does not divide $k + i$ in J since there is no Gaussian integer $a + ib$ such that $k + i = (a + ib)p$; or equivalently $(k/p) + (1/p)i$ is not a Gaussian integer. Similarly p does not divide $k - i$ in J. Thus $p \mid (k + i)(k - i)$ but p divides neither factor. Hence p is not a prime element of J. Since J is Euclidean (2.11), it follows by 6.8 that p is not an irreducible element of J. Clearly $p \neq 0$ and p is not a unit of J. Therefore p factorizes properly in J. Thus $p = xy$ for some $x, y \in J$ where neither x nor y is a unit of J. Hence $N(p) = N(xy)$ and so $p^2 = N(x)N(y)$. Now $N(x)$ and $N(y)$ are both positive integers, neither of which is 1 (because x and y are not units of J). Hence $N(x)N(y) = p^2$ implies that $N(x) = N(y) = p$. We have $x = a + ib$ for some $a, b \in \mathbb{Z}$. Therefore $p = N(x) = a^2 + b^2$. ∎

Example 6.16. Let R be the ring of all complex numbers of the form $a + b\sqrt{(-3)}$ for $a, b \in \mathbb{Z}$. Let N be the usual norm function on R (Question 7, Exercises 1). We shall show that R is not a UFD. Although R is not Euclidean with respect to N, we can still use N as follows to find the units of R and in some cases to prove that elements of R are irreducible. Let $r \in R$ with $r = a + b\sqrt{(-3)}$ with $a, b \in \mathbb{Z}$. Then $N(r) = a^2 + 3b^2$, so that $N(r)$ is a non-negative integer and $N(r) = 0$ if and only if $r = 0$. If also $s \in R$ with $rs = 1$ then $1 = N(1) = N(rs) = N(r)N(s)$. Since $N(r)$ and $N(s)$ are non-negative integers whose product is 1, we must have $N(r) = 1$. Thus $a^2 + 3b^2 = 1$ which gives $a^2 = 1$ and $b^2 = 0$. Therefore we have $r = \pm 1$. Hence the only possible units of R are 1 and -1 and clearly these are, in fact, units.

We shall show that R is not a UFD by considering the factorizations

$$4 = 2 \cdot 2 = (1 + \sqrt{(-3)})(1 - \sqrt{(-3)}).$$

As the units of R are 1 and -1, the factorization $2 \cdot 2$ is genuinely different from the factorization $(1 + \sqrt{(-3)})(1 - \sqrt{(-3)})$. We must show that $2, 1 + \sqrt{(-3)}$ and $1 - \sqrt{(-3)}$ are irreducible elements of R. Note that the norm of each of these elements is 4. Let $r \in R$ with $N(r) = 4$. We shall show that r is irreducible. Clearly $r \neq 0$ and r is not a unit of R. Suppose that r factorizes property in R. Then $r = xy$ for some $x, y \in R$, where neither x nor y is a unit. From the above we know that $N(x) \neq 1$ and $N(y) \neq 1$. However, we also have $4 = N(r) = N(xy) = N(x)N(y)$. Therefore $N(x) = 2 = N(y)$. Now $x = c + d\sqrt{(-3)}$ for some $c, d \in \mathbb{Z}$ so that $N(x) = c^2 + 3d^2$. Hence $c^2 + 3d^2 = 2$, which is impossible. Thus R is not a UFD.

There is a ring S constructed below which is a UFD and which contains R. As in Question 10, Exercises 2 set $w = (1 + \sqrt{(-3)})/2$ and let S be the subring of $\mathbb{C}$ consisting of all $a + bw$ with $a, b \in \mathbb{Z}$. For $a, b \in \mathbb{Z}$, we have $a + b\sqrt{(-3)} = (a - b) + 2bw$ with $a - b, 2b \in \mathbb{Z}$. Hence $R \subseteq S$. We know that S is Euclidean and thus a UFD. In S we can write $4 = 2 \cdot 2 = (1 + \sqrt{(-3)})(1 - \sqrt{(-3)})$. Why does this not contradict the fact that S is a UFD? The answer lies in the fact that S has more units than R. In particular, $u = (1 + \sqrt{(-3)})/2$ is a unit of S. We have $2u = (1 + \sqrt{(-3)})$, so that in S we can change from 2 to $1 + \sqrt{(-3)}$ by multiplying by the unit u of S. Similarly $2v = 1 - \sqrt{(-3)}$ where $v = (1 - \sqrt{(-3)})/2$ is a unit of S. Thus in S the factorizations $2 \cdot 2$ and $(1 + \sqrt{(-3)})(1 - \sqrt{(-3)})$ are in effect the same because $(1 + \sqrt{(-3)})(1 - \sqrt{(-3)}) = 2u \cdot 2v$ where u and v are units of S.

The ring R also shows that HCFs do not necessarily exist. In R the number 2 is a common factor of 4 and $2 + 2\sqrt{(-3)}$. Also $1 + \sqrt{(-3)}$ is a common factor of 4 and $2 + 2\sqrt{(-3)}$. There is no HCF since both 2 and $1 + \sqrt{(-3)}$ are irreducible and neither divides the other. ∎

Exercises 6

1. Show that S is a subring of R in 6.6

2. Show that every prime element of an integral domain is irreducible.

3. Show that an irreducible element of a UFD is prime.

4. Two elements a and b of an integral domain R are said to be *associates* if $a = ub$ for some unit $u \in R$. Prove the following statements:
   (i) 'Being associates' is an equivalence relation on the elements of R.
   (ii) a is an associate of b if and only if $a \mid b$ and $b \mid a$.
   (iii) If $a \mid b$ then every associate of a divides b.
   (iv) If g and h are HCFs of a and b then g is an associate of h and every associate of h is an HCF of a and b.

5. Use Euler's criterion to determine whether there is an integer x such that $7 \equiv x^2 \pmod{31}$.

6. In the context of polynomials in X with integer coefficients, show that the HCF of 2 and X is 1 and that there are no polynomials $s(X)$ and $t(X)$ such that $2s(X) + Xt(X) = 1$.

7. Let R be the ring of all $a + b\sqrt{5}$ where $a, b \in \mathbb{Z}$. Show that R is not a UFD. (*Hint*: $4 = 2 \cdot 2 = (\sqrt{5} + 1)(\sqrt{5} - 1)$.)

8. Let $f(X)$ be a polynomial with real coefficients and let z be a non-real complex number such that $f(z) = 0$. Show that $f(\bar{z}) = 0$ where $\bar{z}$ denotes the complex conjugate of z. Hence show that $f(X) = (X - z)(X - \bar{z})g(X)$ for some polynomial $g(X)$ with real coefficients, and that $(X - z)(X - \bar{z})$ has real coefficients when written as a polynomial in X.

   Hence or otherwise show that if $f(X)$ is an irreducible element of the ring of polynomials in X with real coefficients then $f(X)$ is linear or quadratic.

Part of what is proved in this question is sometimes expressed as 'The roots of a real polynomial equation are real or form complex-conjugate pairs'.

9. Let J be the ring of Gaussian integers and N the usual norm function on J.
   (a) Show that $1 + i$, $1 - i$, $-1 + i$, $-1 - i$ are associates of each other in J and that they are the only Gaussian integers r such that $N(r) = 2$.
   (b) Let r be a common factor of $s, t \in J$. Show that $N(r)$ is a common factor of $N(s)$ and $N(t)$ in $\mathbb{Z}$.
   (c) Find the HCF of $1 + 5i$ and $4 + 2i$ in J.
   (d) Find non-zero elements a and b of J with HCF h such that $N(h)$ is not the HCF of $N(a)$ and $N(b)$.
       Contrast this with (b) which shows that $N(h)$ is a common factor of $N(a)$ and $N(b)$.

10. (The two squares theorem) Let n be a positive integer. Show that n is the sum of two squares if and only if every prime factor of p of n which is congruent to 3 (mod 4) occurs as an even power, i.e. if k is the largest positive integer such that $p^k \mid n$ then k is even.
    (*Hint*: (i) Question 6, Exercises 3, (ii) the product of two integers each of which is a sum of two squares is itself a sum of two squares, (iii) Theorem 6.15).

7 Field of quotients of an integral domain

Let R be an integral domain. We shall show that there is a field F containing R as a subring such that every element of F is a fraction formed from the elements of R. This generalizes the way in which every element of the field $\mathbb{Q}$ is a fraction of elements of $\mathbb{Z}$. The method which we shall give for constructing F from R is rather complicated, but the important point is that it always works. Also, in many cases, given a particular integral domain R, it is easy to find or describe the corresponding field F as will be seen from examples later. In Chapter 8 we shall need the existence of such a field F when R is an arbitrary UFD.

The construction given below is motivated by the following consideration. The rational number $2/3$ can be constructed as a pair $(2, 3)$. However, we have to arrange things so that the pairs $(2, 3)$ and $(10, 15)$ give the same fraction since the fractions $2/3$ and $10/15$ are equal.

Construction 7.1. Let R be an integral domain and let S be the set of all ordered pairs (a, b) with $a, b \in R$ and $b \neq 0$. Eventually (a, b) will correspond to a/b. However, we first set up a procedure which identifies the ordered pairs which give the same fraction.

We define two ordered pairs (a, b) and (a', b') to be *equivalent* if and only if $ab' = a'b$. We denote this relation by $(a, b) \sim (a', b')$. It is easy to check that this is an equivalence relation on S. Let a/b denote the equivalence class of (a, b) and let $F = \{a/b \mid (a, b) \in S\}$. Note that an element of F is actually an equivalence class which can be expressed in more than one way. Thus if $(a, b) \sim (a', b')$ in S then a/b and a'/b' are the same fraction. We shall define addition and multiplication on F in such a way that F becomes a field. Let $a/b, c/d \in F$. Define

$$a/b + c/d = (ad + bc)/bd \quad \text{and} \quad (a/b)(c/d) = ac/bd.$$

The denominator bd is non-zero since both b and d are non-zero. Before we go any further, we must check that the definitions are consistent, that is they are independent of the representatives chosen. We shall do this for the addition. Let $a/b = a'/b'$ and $c/d = c'/d'$. We must show that

$$(ad + bc)/bd = (a'd' + b'c')/b'd'.$$

We have $(a, b) \sim (a', b')$ and $(c, d) \sim (c', d')$. So $ab' = a'b$ and $cd' = c'd$. Hence

$$(ad + bc)b'd' = adb'd' + bcb'd' = a'dbd' + bc'b'd = (a'd' + b'c')bd.$$

Thus $(ad + bc, bd) \sim (a'd' + b'c', b'd')$ as required. The argument for the consistency of the multiplication is similar. It is straightforward but somewhat tedious to show that F is a ring. The zero element of F is $0/1$ and the multiplicative identity is $1/1$. (These are usually denoted simply by 0 and 1 respectively.) If a/b is a non-zero element of F with $a, b \in R$ then it is easy to see that $a \neq 0$. Hence $b/a \in F$ and we have $(a/b)(b/a) = ab/ab = 1$. Thus a/b has inverse b/a and F is a field. Strictly speaking R is not a subset of F. We get round this problem by identifying $a \in R$ with $a/1 \in F$. After this identification R can be considered a subring of F and a typical element of F is expressible as $a/b = (a/1)(b/1)^{-1}$.   ■

Definition 7.2. Let R be an integral domain. A field F is called the *quotient field* or the *field of quotients* or the *field of fractions* of R if R is a subring of F and every element of F is expressible as ab^{-1} where $a, b \in R$ and $b \neq 0$.

Comments 7.3

(i) We call F 'the' quotient field since it can be shown that any two fields satisfying Definition 7.2 'look alike', i.e. they are isomorphic (13.7).

(ii) We have shown in 7.1 that every integral domain has a field of fractions.

Example 7.4. Let R be the ring of polynomials in X over the field F. The quotient field of R is denoted $F(X)$ and is called the *field of rational functions in X over F*. The elements of $F(X)$ are of the form $f(X)/g(X)$ where $f(X)$ and $g(X)$ are polynomials in X over F with $g(X) \neq 0$.   ■

Example 7.5. Let J be the ring of Gaussian integers and let F be the set of complex numbers of the form $x + iy$ with $x, y \in \mathbb{Q}$. It is easy to check that F is a subring of $\mathbb{C}$. Let $x + iy$ be a non-zero element of F. Not both x and y are zero, so that $x^2 + y^2 \neq 0$. We have

$$1/(x + iy) = (x - iy)/(x^2 + y^2) = x/(x^2 + y^2) - iy/(x^2 + y^2),$$

where $x/(x^2 + y^2)$ and $-y/(x^2 + y^2) \in \mathbb{Q}$. Thus $1/(x + iy) \in F$ and F is a field. We shall show that F is the quotient field of J. Let $x + iy \in F$. We must write $x + iy$ as a fraction of two elements of J, i.e. as a fraction of two complex numbers with integer coordinates. We can write $x = a/d$ and $y = b/d$ for some $a, b, d \in \mathbb{Z}$ with $d \neq 0$. Thus $x + iy = (a + ib)d^{-1}$ with $a + ib, d \in J$. Hence F is the quotient field of J. This field F is called the *field of Gaussian numbers*.   ■

Example 7.6. Let F be a field and let R be the subring of $F[X]$ consisting of those polynomials with 0 as coefficient of X. We already know a field in this context, namely the field $F(X)$ constructed in 7.4. We shall show that $F(X)$ is also the quotient field of R. Let $a \in F(X)$. We must show that a is a fraction of two elements of R. Certainly we have $a = f(X)/g(X)$ for some $f(X), g(X) \in F[X]$ with $g(X) \neq 0$. However, $f(X)$ or $g(X)$ may not belong to R, i.e. the coefficient of X may be non-zero. To avoid this problem we write $a = X^2 f(X)/X^2 g(X)$ where $X^2 f(X)$, $X^2 g(X) \in R$. Therefore every element of the field $F(X)$ is a fraction of elements of R and so $F(X)$ is the quotient field of R.   ■

Exercises 7

1. Let F be the set of all real numbers of the form $a + b\sqrt{2}$ for $a, b \in \mathbb{Q}$. Prove that F is a field. Let R be the subring of F consisting of all $a + b\sqrt{2}$ for $a, b \in \mathbb{Z}$. Show that F is the quotient field of R.

2. Prove that $\mathbb{Q}(X)$ is the quotient field of $\mathbb{Z}[X]$.

3. Let $f(X)$ be a non-zero polynomial of degree n with coefficients in an integral domain R. Show that the equation $f(X) = 0$ has at most n roots in R. (Compare this with Question 9, Exercises 1 where there is a quadratic equation with 4 roots.)

4. Let R be an integral domain with quotient field F and let $x_1, \ldots, x_n$ be a finite number of elements of F. Show that there are elements $a_1, \ldots, a_n$ and b of R such that $x_i = a_i/b$ for all i. (In other words, any finite number of elements of the quotient field of R can be expressed as fractions with a common denominator.)

8   Factorization of polynomials

In order to motivate the material of this chapter, consider the following simple-looking problem: Set $f(X) = 8X^9 + 56X^4 + 49X + 28$; can $f(X)$ be written as the product of two polynomials of lower degree with rational number coefficients? The most direct approach would seem to be to suppose that $f(X)$ does factorize, equate it to the product of two factors with unknown coefficients, multiply the factors together and equate coefficients to get a set of equations for the unknown coefficients, solve these equations or show that they have no solution; but $f(X)$ might be the product of polynomials of degrees 1 and 8, or 2 and 7, or 3 and 6, or 4 and 5; it is starting to look very complicated and messy.

In later parts of this book we shall need to be able to answer this sort of question, preferably without doing too much work. But it is also the sort of elementary question which keeps occurring in other and more advanced topics such as Galois theory and the unsolvability of the quintic, algebraic number theory and so on. If possible, we want to develop systematic techniques for tackling such questions. The nature of the problem depends critically on the sort of coefficients which are being used. For example, we can ask questions similar to the one above but taking the coefficients of the given polynomial and its possible factors to be in the field of integers modulo p for some prime number p. Although there are some general underlying principles, the techniques for solving these problems tend to vary according to the type of coefficient. For instance, once we have proved the Eisenstein criterion it will be possible to answer the question above about $f(X)$ in one line (come back later and try it), but that technique does not always work for polynomials with rational number coefficients, and it is of no use for polynomials over the integers modulo p.

Many of the results of this chapter should be regarded as being problem-solving techniques even though they are called theorems, and this will become clear from the worked examples. On the other hand, there is also the very important theoretical question of the uniqueness of factorization of polynomials. We shall prove in particular that the ring of polynomials in X with integer coefficients ia a unique factorization domain. Not only does this establish the uniqueness of the answer to factorization questions over the integers, but it also gives an important example of a UFD which is not Euclidean (so far, the only way in which we have been able to prove the unique factorization property has been via the very special context of Euclidean domains).

The main theorem of this chapter will be that if R is a *UFD* then so is the ring $R[X]$. This shows in particular that $\mathbb{Z}[X]$ and $F[X, Y]$, for any field F and independent indeterminates X and Y, are UFDs. So far the only examples we have given of UFDs have been Euclidean rings. The above shows that there are many UFDs which are not Euclidean.

Comment 8.1. In questions concerning factorization of polynomials, it is crucial to understand precisely what is meant by a proper factorization. It does not necessarily mean factorization as a product of polynomials of lower degree. For example, $2 \cdot X$ is a proper factorization of $2X$ in $\mathbb{Z}[X]$ because neither 2 nor X is a unit of $\mathbb{Z}[X]$. On the other hand $2 \cdot X$ is not a proper factorization of $2X$ in $\mathbb{Q}[X]$ since 2 is a unit of $\mathbb{Q}[X]$ (its inverse of course being $1/2$). For any field F, a proper factorization in $F[X]$ does mean a factorization by polynomials of lower degree, because the only trivial factors (i.e. units) are the non-zero constants. However, over a ring such as $\mathbb{Z}$ which is not a field, we must make a careful note of what the *trivial factors* (i.e. units) are, and this will depend on the choice of R. Thus, for example, the only units of $\mathbb{Z}[X]$ are 1 and -1.

Definition 8.2. Let R be an integral domain with quotient field F. An element $f(X)$ of $F[X]$ is said to be *primitive* if the coefficients of $f(X)$ belong to R and have no non-trivial common factor in R.

Example 8.3. Take $R = \mathbb{Z}$ and $F = \mathbb{Q}$. Then $2X^5 - 4X^2 + 3$ is primitive because its coefficients are in $\mathbb{Z}$ and the only common factors of 2, -4 and 3 in $\mathbb{Z}$ are the units 1 and -1. On the other hand $2X^5 - 4X^2 + 3/2$ is not primitive because its coefficients are not all in $\mathbb{Z}$ and $2X^4 - 4X^2 + 4$ is not primitive because its coefficients have a non-trivial common factor, i.e. 2. ∎

Lemma 8.4 (Gauss's lemma). *Let R be a UFD with quotient field F. Then the product of any two primitive elements in $F[X]$ is primitive.*

Proof. Let $f(X)$ and $g(X)$ be two primitive elements of $F[X]$. We have $f(X) = a_0 + a_1 X + \cdots + a_n X^n$ for some $a_i \in R$ where the a_i have no non-trivial common factor. Similarly $g(X) = b_0 + b_1 X + \cdots + b_k X^k$ for some $b_i \in R$ where the b_i have no non-trivial common factors. We have

$$f(X)g(X) = c_0 + c_1 X + \cdots + c_{n+k} X^{n+k}$$

where $c_i = a_0 b_i + a_1 b_{i-1} + \cdots + a_{i-1} b_1 + a_i b_0$. Here we set $a_j = 0$ for all $j > n$ and $b_j = 0$ for all $j > k$. Clearly $c_i \in R$ for all i.

Let p be an irreducible element of R. We shall show that p does not divide all the c_i. It follows from this that the c_i have no non-trivial common factor. Since $f(X)$ is primitive, not all the a_i are divisible by p. We go through the list $a_0, a_1, \ldots$ until we find the first one which is not divisible by p. Thus there is an integer s such that $p \mid a_i$ for all $i < s$ but p does not divide a_s. Similarly there is an integer t such that $p \mid b_i$ for all $i < t$ but p does not divide b_t. We have

$$c_{s+t} = a_0 b_{s+t} + a_1 b_{s+t-1} + \cdots + a_{s-1} b_{t+1} + a_s b_t + \cdots + a_{s+t} b_0. \quad (*)$$

All the terms on the right-hand side of $(*)$ are divisible by p except possibly for $a_s b_t$. Now p divides neither a_s nor b_t. Since R is UFD, p is a prime element (Question 3, Exercise 6). Hence p does not divide $a_s b_t$. Therefore p divides all but one of the terms on the right-hand side of $(*)$. Thus p does not divide c_{s+t}. ∎

Lemma 8.5. *Let R be a UFD and let $a_1, \ldots, a_n \in R$ such that not all a_i are zero. Then there are elements $h, b_1, \ldots, b_n \in R$ such that $a_i = hb_i$ for all i, and no non-trivial element of R divides all the b_i. In particular, h is an HCF of the a_i.*

Proof. Since R is a UFD every non-zero element of R is expressible uniquely as a product of irreducible elements. Let $p_1, \ldots, p_k$ be the distinct irreducible elements of R such that each p_i is a factor of at least one non-zero a_j. For each i from 1 to k let $h(i)$ be the largest non-negative integer such that $p_i^{h(i)} \mid a_j$ for all j. In other words, we take $p_i^{h(i)}$ to be the largest power of p_i which divides all the a_j. Set $h = p_1^{h(1)} p_2^{h(2)} \cdots p_k^{h(k)}$. Then h is an HCF of the a_j. Now set $b_i = a_i/h$ for all i. Then $b_i \in R$ since $h \mid a_i$. As the b_j are formed from the a_j by dividing out their HCF, it follows that the b_j have no non-trivial common factors. ∎

Lemma 8.6. *Let R be a UFD with quotient field F and let $f(X)$ be a non-zero element of $F[X]$. Then $f(X) = c_f f^*(X)$ for some primitive element $f^*(X)$ of $F[X]$ and some $c_f \in F$. Also c_f and $f^*(X)$ are unique except for multiplication by units of R.*

Proof. (First we consider an example. Suppose that $R = \mathbb{Z}$ and $F = \mathbb{Q}$. Let $f(X) = (3/4)X^2 - 6/5$. We put the coefficients over a common denominator so that $f(X) = (15/20)X^2 - 24/20$. Now we find the HCF of the numerators 15 and 24 of the coefficients, which is 3. We write $f(X) = (3/20)(5X^2 - 8)$ so that we can take $c_f = 3/20$ and $f^*(X) = 5X^2 - 8$.) In general we write the coefficients of $f(X)$ as fractions of elements of R with a common denominator (see Question 4, Exercises 7). Thus $f(X) = (a_0 + a_1X + \cdots + a_nX^n)/d$ for some a_i, $d \in R$ with $a_n \neq 0$ and $d \neq 0$. As in Lemma 8.5 let b be an HCF of the a_i in R with $a_i = bc_i$. Then $f(X) = c_f f^*(X)$ where $c_f = b/d$ and $f^*(X) = c_0 + c_1X + \cdots + c_nX^n$. The c_i have no non-trivial common factors so that $f^*(X)$ is primitive.

For uniqueness suppose that $f(X) = c_f f^*(X) = qg(X)$ for some primitive element $g(X) \in F[X]$ and some $q \in F$. We have $g(x) = (c_f/q)f^*(X)$. Since $c_f/q \in F$ we have $c_f/q = u/v$ for some $u, v \in R$ with $v \neq 0$. Thus $vg(X) = uf^*(X)$. Since R is a UFD, just as when writing rational numbers in their lowest terms, we may suppose that u and v have no non-trivial common factor. Suppose that v is not a unit of R. Then v has an irreducible factor p. Since p divides all the coefficients of $vg(X)$ it also divides all the coefficients of $uf^*(X)$. However, since R is a UFD, p is a prime element. Now u and v are relatively prime and p divides v, so that p does not divide u. Hence p must divide all the coefficients of $f^*(X)$ which is a contradiction. Thus v is a unit of R. Similarly u is a unit of R. Set $w = v/u$. Then w is a unit of R. Thus $q = wc_f$ and $g(X) = w^{-1}f^*(X)$ as required.   ∎

Definition 8.7. c_f and $f^*(X)$ as in 8.6 are called the *content* and the *primitive* of $f(X)$.

Lemma 8.8. *Let R be a UFD with quotient field F. Let $g(X), h(X)$ be non-zero elements of $F[X]$. Set $f(X) = g(X)h(X)$ and write $f(X) = c_f f^*(X)$, $g(X) = c_g g^*(X)$, $h(X) = c_h h^*(X)$ as in 8.6. Then there is a unit $u \in R$ such that $c_f = uc_g c_h$ and $f^*(X) = u^{-1}g^*(X)h^*(X)$.*

Proof. We have $c_f f^*(X) = c_g c_h g^*(X)h^*(X)$ where $g^*(X)h^*(X)$ is primitive by 8.4. The result now follows immediately by the uniqueness proved in 8.6.   ∎

Theorem 8.9. *Let R be a UFD with quotient field F and let $f(X) \in R[X]$ with $f(X) \neq 0$. Suppose that we have $f(X) = g(X)h(X)$ for some $g(X)$, $h(X) \in F[X]$. Then $f(X) = a(X)b(X)$ for some $a(X), b(X) \in R[X]$ with $\partial(a) = \partial(g)$ and $\partial(b) = \partial(h)$.*

(The practical importance of this is that if $f(X)$ has coefficients in $\mathbb{Z}$ and we are investigating whether $f(X)$ factorizes properly over $\mathbb{Q}$, then we need only consider factorizations of $f(X)$ into products of polynomials with coefficients in $\mathbb{Z}$. Thus if $10X^5 + X^3 + 3X + 144$ is a product of a quadratic and a cubic with coefficients in $\mathbb{Q}$ then it is also the product of a quadratic and a cubic with coefficients in $\mathbb{Z}$.)

Proof. We have $f(X) = c_f f^*(X) = c_g c_h g^*(X) h^*(X)$ in the usual notation. By 8.8 we have $c_g c_h = u c_f$ for some unit u of R. Since $f(X)$ has
coefficients in R, we can take c_f to be an HCF of the coefficients of $f(X)$.
Thus $c_f \in R$ so that $c_g c_h \in R$. Take $a(X) = c_g c_h g^*(X)$ and $b(X) = h^*(X)$ ∎

Example 8.10. Set $f(X) = 8X^3 - 6X - 1$. We shall investigate whether
$f(X)$ is irreducible over $\mathbb{Q}$, i.e. whether $f(X)$ is an irreducible element of
$\mathbb{Q}(X)$. Since the coefficients of $f(X)$ are in $\mathbb{Z}$, by 8.9 we need only
investigate whether $f(X)$ factorizes as a product of polynomials of lower
degree with coefficients in $\mathbb{Z}$. Any such factorization would have to be
'linear times quadratic' because this includes the case 'linear times linear
times linear'. Can we have $8X^3 - 6X - 1 = (aX + b)(cX^2 + dX + e)$ with
$a, b, c, d, e \in \mathbb{Z}$? Equating coefficients gives $ac = 8$, $bc + ad = 0$, $ae + bd =
-6$ and $be = -1$. Now because $be = -1$ and b and e *are integers* we have
either $b = 1$ and $e = -1$ or $b = -1$ and $e = 1$. Also $ac = 8$ so that a is one
of the numbers 1, 2, 4, 8, -1, -2, -4, -8. Suppose that $a = 8$ and $b = 1$.
Then $8X + 1$ is a factor of $f(X)$ so that $f(-1/8) = 0$. However, direct
calculation shows that $f(-1/8) \neq 0$. Similarly all the other possibilities for
$aX + b$ can be ruled out by direct calculation. Therefore $f(X)$ is irreducible over $\mathbb{Q}$.   ∎

Example 8.11. Set $h(X) = X^4 - 2X^2 + 9$. We shall investigate whether
$h(X)$ factorizes as a product of two quadratics over $\mathbb{Q}$. By 8.9 we need only
consider the possibility that $h(X) = (aX^2 + bX + c)(dX^2 + eX + f)$ for
$a, b, c, d, e, f \in \mathbb{Z}$. Equating coefficients of X^4 gives $ad = 1$ where a and d
are integers. Hence either $a = d = 1$ or $a = d = -1$. In the second case we
can multiply each of the quadratics by -1 so that without loss of generality
we may assume $a = d = 1$. Equating coefficients above gives $b + e = 0$,
$c + f + be = -2$, $bf + ce = 0$ and $cf = 9$. Since $e = -b$ we have $0 =
bf + ce = b(f - c)$, so that either $b = 0$ or $c = f$. Now as $cf = 9$ we know
that c and f are both positive or both negative. Suppose that $b = 0$. Then
$-2 = c + f + be = c + f$. Since c and f have the same sign and $c + f = -2$
we must have $c = f = -1$, which contradicts $cf = 9$. Therefore $b \neq 0$ so that
$c = f$. Now $cf = 9$ implies that either $c = f = 3$ or $c = f = -3$. However,
$c + f + be = -2$. Therefore $b^2 = -be = c + f + 2$ so that either $b^2 = 8$ or
$b^2 = -4$, both of which are contradictions. Hence $h(X)$ is not a product of
two quadratics over $\mathbb{Q}$.   ∎

Theorem 8.12. *Let R be a UFD. Then $R[X]$ is a UFD.*

Proof. Let $f(X)$ be a non-trivial element of $R[X]$. We shall first show that $f(X)$ is a product of irreducible elements of $R[X]$. By 8.5 and 8.6 we have $f(X) = c_f f^*(X)$ where c_f is an HCF of the coefficients of $f(X)$ and $f^*(X)$ is primitive. In R we have $c_f = p_1 p_2, \ldots, p_s$ where each p_i is an irreducible element of R. (If c_f is a unit of R we can forget it). Each p_i is irreducible as an element of $R[X]$ (see Question 6, Exercises 8). Suppose that $f^*(X) = g(X)h(X)$ with $g(X), h(X) \in R[X]$. If $g(X)$ is a constant polynomial then the element $g(X)$ of R divides all the coefficients of $f^*(X)$, so that $g(X)$ is a unit of R. It follows that if $f^*(X) = g(X)h(X)$ is a proper factorization of $f^*(X)$ in $R[X]$, then $g(X)$ and $h(X)$ have lower degree than $f^*(X)$. Also $g(X)$ and $h(X)$ are primitive (see Question 6, Exercises 8). By repeatedly factorizing into factors of lower degree we obtain $f^*(X) = q_1(X)q_2(X), \ldots, q_t(X)$ where each $q_i(X)$ is a primitive irreducible element of $R[X]$. Thus $f(X) = p_1 p_2 \cdots p_s q_1(X)q_2(X) \cdots q_t(X)$ where each p_i and $q_j(X)$ is irreducible as an element of $R[X]$.

Let w be an irreducible element of $R[X]$. Then either $\partial(w) = 0$, i.e. $w \in R$ and w is an irreducible element of R; or $\partial(w)$ is positive and the irreducibility of w implies that its coefficients have no common factors in R, so that w is primitive. To prove the uniqueness of the factorization of $f(X)$ into irreducible factors, suppose that

$$f(X) = p_1 p_2 \cdots p_s q_1(X)q_2(X) \cdots q_t(X)$$

$$= a_1 a_2 \cdots a_u b_1(X)b_2(X) \cdots b_v(X)$$

where each p_i and a_j is an irreducible element of R and each $q_m(X)$ and $b_n(X)$ is a primitive irreducible element of $R[X]$. By 8.6 the content of $f(X)$ is unique and so we have $a_1 a_2 \cdots a_u = z p_1 p_2 \cdots p_s$ for some unit z of R. Since R is UFD it follows that $u = s$ and that the a_is are associates of the p_is in some order. We have

$$q_1(X)q_2(X) \cdots q_t(X) = z b_1(X)b_2(X) \cdots b_v(X).$$

Let F be the quotient field of R. Note that z is a unit of $F[X]$. Now each $q_i(X)$ and $b_j(X)$ is irreducible as an element of R[X] and hence by 8.9 also as an element of $F[X]$. However, $F[X]$ is a Euclidean ring and hence a UFD (2.9 and 6.12). Therefore $v = t$ and by renumbering if necessary we have $q_i(X) = z_i b_i(X)$ for all i where each z_i is a unit of $F[X]$, i.e. z_i is a non-zero element of F. Now the content of $q_i(X)$ is 1 and the content of $z_i b_i(X)$ is z_i. Therefore by 8.6 each z_i is a unit of R. Thus for each i we have $q_i(X) = z_i b_i(X)$ for some unit z_i of R.

Thus, for example, if $R = \mathbb{Z}$ and $f(X) = 24X^3 + 24X^2 + 12X + 12$ then $f(X) = 2 \cdot 2 \cdot 3 \cdot (X + 1)(2X^2 + 1)$ where $2, 3, X + 1, 2X^2 + 1$ are irreducible elements of $\mathbb{Z}[X]$.

Corollary 8.13. *If R is a UFD then so is the ring $R[X_1, X_2, \ldots, X_n]$, where $X_1, \ldots, X_n$ are independent indeterminates.*

Proof. We shall proceed by induction on n. When $n = 1$ we know that $R[X_1]$ is a UFD by 8.12. In general we have $R[X_1, X_2, \ldots, X_n] = S[X_n]$ where $S = R[X_1, \ldots, X_{n-1}]$. By the induction hypothesis S is a UFD. So again by 8.12 $S[X_n] = R[X_1, X_2, \ldots, X_n]$ is a UFD.   ∎

As a special case of the above we have

Corollary 8.14. *(i) $\mathbb{Z}[X]$ is a UFD. (ii) If F is a field then $F[X_1, X_2, \ldots, X_n]$ is a UFD.*   ∎

Example 8.15. In $\mathbb{Z}[X]$ the unique factorization of $6X^2 + 18X + 12$ into irreducible factors is $2 \cdot 3 \cdot (X + 1) \cdot (X + 2)$. Note that 2 and 3 are irreducible elements of $\mathbb{Z}$, and that $X + 1$ and $X + 2$ are primitive irreducible elements of positive degree of $\mathbb{Z}[X]$.   ∎

Comments 8.16

(a) The remainder of this chapter consists of results and examples concerned with the problems of determining whether a polynomial is irreducible and of writing a given polynomial as a product of irreducible ones. The nature of the problems depends critically on what sort of coefficients are allowed.

(b) In $\mathbb{C}[X]$ a polynomial is irreducible if and only if it is linear. The problem of factorizing an element $f(X)$ of $\mathbb{C}[X]$ into irreducible polynomials is equivalent to finding all the roots of the equation $f(X) = 0$, which may be a very difficult problem if $\partial(f)$ is large or if the coefficients and/or roots are not convenient numbers.

(c) We know from Question 8, Exercises 6 that a polynomial over $\mathbb{R}$ is irreducible if and only if it is linear or of the form $aX^2 + bX + c$ with $a, b, c \in \mathbb{R}$ where the quadratic equation $aX^2 + bX + c = 0$ has no real roots. Working over $\mathbb{R}$, the problem of factorizing $f(X)$ into irreducible polynomials corresponds to finding the real roots and the complex conjugate pairs of non-real roots of the equation $f(X) = 0$. The real roots give linear factors and the complex conjugate pairs correspond to irreducible quadratic factors.

(d) These problems are much harder over the rational numbers $\mathbb{Q}$. There are, for example, irreducible polynomials of arbitrary large degree over $\mathbb{Q}$. However polynomials over $\mathbb{Q}$ play a crucial role in number theory and other applications. If $f(X)$ has coefficients in $\mathbb{Q}$, there is usually no harm in multiplying through by an integer to get rid of denominators of the coefficients. Thus we have a polynomial $f(X)$ with coefficients in $\mathbb{Z}$ and we are concerned with factorizing it over $\mathbb{Q}$. We have two main facts to help us: By 8.9 we need only look for possible factorizations of $f(X)$ into factors with coefficients in $\mathbb{Z}$ and we may be able to use either the method given in (e) or the Eisenstein criterion below to show that $f(X)$ is irreducible.

(e) Sometimes we can make use of the fact that if $f(X)$ factorizes over $\mathbb{Z}$ then there is a corresponding factorization over $\mathbb{Z}/p\mathbb{Z}$ for every prime p. For example if over $\mathbb{Z}$ we have

$$f(X) = (22X^2 - 6X + 30)(15X^3 + 10),$$

then over $\mathbb{Z}/7\mathbb{Z}$ we have

$$f(X) = (X^2 + X + 2)(X^3 + 3),$$

where we have replaced all the coefficients by the corresponding elements of $\mathbb{Z}/7\mathbb{Z}$. To avoid technical problems, it is best only to use a prime p which does not divide the leading coefficient of $f(X)$ (otherwise the degree may drop). For example, if $f(X)$ is as above then over $\mathbb{Z}/5\mathbb{Z}$ we have

$$f(X) = (2X^2 - X + 0)(0X^3 + 0) = 0,$$

which is true but not of much use.

The point of these remarks is: let $f(X) \in \mathbb{Z}[X]$ and suppose that there is a prime number p such that p does not divide the leading coefficient of $f(X)$ and $f(X)$ is irreducible over $\mathbb{Z}/p\mathbb{Z}$. Then $f(X)$ is not the product over $\mathbb{Z}$ of two polynomials of lower degree (but $f(X)$ may still have an obvious factorization such as $4X + 2 = 2(2X + 1)$).

(f) Recall that if $f(X)$ is a polynomial with coefficients in a field F and if $a \in F$ then $X - a$ is a factor of $f(X)$ if and only if $f(a) = 0$. Thus if $F = \mathbb{Z}/p\mathbb{Z}$ with p prime, we can look for a linear factor $X - a$ of $f(X)$ over F by considering each element $a \in \mathbb{Z}/p\mathbb{Z}$ and determining whether $f(a) = 0$.

Theorem 8.17 (Eisenstein criterion). *Let R by a UFD with quotient field F and let $f(X)$ be a non-constant polynomial with coefficients in R. Suppose that there is an irreducible element $p \in R$ such that*

(i) *p does not divide the leading coefficient of $f(X)$,*

(ii) *p divides all the other coefficients of $f(X)$, and*

(iii) *p^2 does not divide the constant term of $f(X)$.*

Then $f(X)$ is an irreducible element of $F[X]$.

Proof. We have $f(X) = a_0 + a_1 X + \cdots + a_n X^n$ for some $a_i \in R$ with $a_n \neq 0$. By 8.9 we need only consider possible factorizations of $f(X)$ into factors with coefficients in R. Suppose that $f(X) = g(X)h(X)$ for some $g(X), h(X) \in R[X]$. We have $g(X) = b_0 + b_1 X + \cdots + b_s X^s$ and $h(X) = c_0 + c_1 X + \cdots + c_t X^t$ for some $b_i, c_j \in R$ with $b_s \neq 0 \neq c_t$. We have $s + t = n$. Equating constant terms and leading coefficients of $f(X)$ and $g(X)h(X)$ gives $a_0 = b_0 c_0$ and $a_n = b_s c_t$. Since p divides a_0 it follows that p divides b_0 or c_0. Without loss of generality we suppose that p divides b_0. Since p^2 does not divide a_0 it follows that p does not divide c_0. Also p does not divide a_n so that p does not divide b_s. Thus p divides b_0 but p does not divide b_s. Let i be the smallest positive integer such that p does not divide b_i. Equating the coefficients of $f(X)$ and $g(X)h(X)$ gives

$$a_i = b_0 c_i + b_1 c_{i-1} + \cdots + b_{i-1} c_1 + b_i c_0. \qquad (*)$$

By definition of i we know that p divides b_j for all $j < i$. Thus p divides every term on the right-hand side of $(*)$ except possibly $b_i c_0$. However, p divides neither b_i nor c_0 and so since p is prime, p does not divide $b_i c_0$. Hence p does not divide the right-hand side of $(*)$. Therefore p does not divide a_i. Hence $i = n$. However, $n = s + t$ and $i \leq s$. Therefore $s = n$ and $t = 0$, so that $h(X)$ is a constant polynomial. Thus the only factorizations of $f(X)$ in $R[X]$ are of the form $f(X) = g(X) \cdot h$ for some $h \in R$. It follows by 8.9 that $f(X)$ has no proper factorization in $F[X]$. ■

Example 8.18. $X^9 + 4X^2 + 6$ is irreducible over $\mathbb{Q}$ by the Eisenstein criterion with $p = 2$ because 2 does not divide the leading coefficient 1; 2 divides all other coefficients which are 0, 4 or 6; and 4 does not divide the constant term 6. ■

Example 8.19. $X + 1$ is irreducible over $\mathbb{Q}$ but there is no prime p which would enable us to apply the Eisenstein criterion. ■

Definition 8.20. A polynomial is said to be *monic* if the coefficient of the leading term is 1.

Proposition 8.21. *Let R be a UFD with quotient field F and let $f(X)$ be a monic polynomial with coefficients in R. Suppose that $f(a) = 0$ for some $a \in F$. Then $a \in R$ and a divides the constant term of $f(X)$.*

Proof. We have $f(X) = r_0 + r_1 X + \cdots + r_{n-1} X^{n-1} + X^n$ for some $r_i \in R$. Also $a = b/c$ for some $b, c \in R$ with $c \neq 0$ and $(b, c) = 1$. We have

$$0 = f(a) = f(b/c) = (b/c)^n + \cdots + r_1(b/c) + r_0.$$

Hence

$$b^n = -(r_{n-1} b^{n-1} c + \cdots + r_1 b c^{n-1} + r_0 c^n). \qquad (*)$$

Suppose that c is not a unit of R. Let p be an irreducible factor of c. Then p divides every term on the right-hand side of $(*)$. Hence p divides b^n so that p divides b. Thus p is a non-trivial common factor of b and c. This is a contradiction. Hence c is a unit of R. Thus $c^{-1} \in R$ and so $a = bc^{-1} \in R$. Also a divides r_0 because

$$r_0 = -(r_1 + \cdots + r_{n-2}a^{n-2} + a^{n-1})a. \qquad \blacksquare$$

Example 8.22. Set $f(X) = X^4 - 2X^2 + 9$. We know by 8.11 that $f(X)$ is not the product of two quadratics over $\mathbb{Q}$. If $f(X)$ has a linear factor $X - a$ over $\mathbb{Q}$ then $f(a) = 0$ and so by 8.21 $a \in \mathbb{Z}$ and a divides the constant term of $f(X)$. Thus a is one of the numbers 9, 3, 1, -1, -3, -9. Direct calculation shows that in each case $f(a) \neq 0$. Therefore $f(X)$ has no quadratic or linear factor over $\mathbb{Q}$ so that $f(X)$ is irreducible over $\mathbb{Q}$. $\blacksquare$

Example 8.23. Let $f(X) = X^4 + 1$ and $F = \mathbb{Z}/3\mathbb{Z}$. Is $f(X)$ irreducible over F? The elements of F are 0, 1, -1, with $-1 = 2$. Clearly there is no $a \in F$ with $f(a) = 0$ so that $f(X)$ has no linear factors over the field F. Suppose that $f(X)$ is the product of two quadratic factors over F. Since F is a field and the leading coefficient of $f(X)$ is 1, we may without loss of generality suppose that $f(X) = (X^2 + aX + b)(X^2 + cX + d)$ with $a, b, c, d \in F$. Equating coefficients gives $a + c = 0$, $b + d + ac = 0$, $ad + bc = 0$ and $bd = 1$. In F the equation $bd = 1$ gives either $b = d = 1$ or $b = d = -1$. Hence $0 = b + d + ac = 2b + ac = 2b - a^2$ so that $a^2 = 2b$. In F we cannot have $a^2 = 2$. Therefore $b \neq 1$ so that $b = -1 = d$. The equations for a, b, c, and d are satisfied if we take $a = 1$ and $c = -1$. In fact over F we have $X^4 + 1 = (X^2 + X - 1)(X^2 - X - 1)$ because $-3X^2 = 0$. $\blacksquare$

Example 8.24. Is $f(X) = X^3 + 2$ irreducible over $\mathbb{Z}/7\mathbb{Z}$? Any proper factorization of $f(X)$ would involve a linear factor $X - a$ for some $a \in \mathbb{Z}/7\mathbb{Z}$. In $\mathbb{Z}/7\mathbb{Z}$ we have $0^3 = 0$, $1^3 = 1$, $2^3 = 8$, $3^3 = 27 = -1$, $4^3 = (-3)^3 = 1$, $5^3 = (-2)^3 = -1$, $6^3 = (-1)^3 = -1$. Thus there is no $a \in \mathbb{Z}/7\mathbb{Z}$ such that $f(a) = 0$. Hence $f(X)$ does not have a factor of the form $X - a$ and so $f(X)$ is irreducible over $\mathbb{Z}/7\mathbb{Z}$. $\blacksquare$

Example 8.25. Is $f(X) = 50X^3 + 49X + 702$ irreducible over $\mathbb{Z}$? The usual method of testing for linear factors would be very tedious. Instead we proceed as in 8.16(e) and reduce mod p for a suitable prime p. We do not want p to divide the leading coefficient 50 so we do not take $p = 2$ or $p = 5$. We could try $p = 3$ but mod 3 we have $f(X) = 2X^3 + X$ which factorizes and this has no bearing on the irreducibility of $f(X)$ over $\mathbb{Z}$. Next we try $p = 7$. Over $\mathbb{Z}/7\mathbb{Z}$ we have $f(X) = X^3 + 2$ which is irreducible as shown in 8.24. Therefore $f(X)$ is not the product over $\mathbb{Z}$ of two polynomials of lower degree. Since the coefficients of $f(X)$ are relatively prime, there is no proper factorization of $f(X)$ over $\mathbb{Z}$.   ∎

Example 8.26.   Over $\mathbb{Z}/11\mathbb{Z}$ we have

$$X^4 + 2 = X^4 - 9 = (X^2 - 3)(X^2 + 3).$$

To determine whether $X^2 - 3$ and $X^2 + 3$ are irreducible over $\mathbb{Z}/11\mathbb{Z}$ we check if the equations $X^2 = 3$ and $X^2 = -3$ have any solutions. In $\mathbb{Z}/11\mathbb{Z}$ it is convenient to write the elements as integers from -5 to 5. We have $0^2 = 0$, $(\pm 1)^2 = 1$, $(\pm 2)^2 = 4$, $(\pm 3)^2 = 9$, $(\pm 4)^2 = 16 = 5$, $(\pm 5)^2 = 25 = 3$. Therefore 5 and -5 are roots of $X^2 - 3 = 0$ in $\mathbb{Z}/11\mathbb{Z}$, and there are no roots of $X^2 + 3 = 0$. Therefore the irreducible factors of $f(X)$ over $\mathbb{Z}/11\mathbb{Z}$ are $X - 5$, $X + 5$ and $X^2 + 3$.   ∎

Sometimes it is helpful to change the indeterminate in order to test for irreducibility as the next Proposition and Example show.

Proposition 8.27.   *Let R be any ring, $a \in R$ and $f(X) \in R[X]$. Set $Y = X - a$. Then $f(X)$ is irreducible as an element of $R[X]$ if and only if $f(Y + a)$ is an irreducible element of $R[Y]$.*

In other words $f(X)$ is irreducible if and only if it is irreducible as a polynomial in Y after the change of variable given by $X = Y + a$.

Proof.   Since $f(X) \in R[X]$, we can rewrite $f(Y + a)$ as a polynomial in Y with coefficients in R. For instance if we have $f(X) = uX^2 + vX + w$ with $u, v, w \in R$ then

$$f(Y + a) = u(Y + a)^2 + v(Y + a) + w = uY^2 + (2ua + v)Y + ua^2 + va + w.$$

Thus, in general, we have $f(X) = g(Y)$ for some $g(Y) \in R[Y]$.

Suppose that $f(X) = p(X)q(X)$ is a proper factorization of $f(X)$ in $R[X]$. Then $f(Y + a) = p(Y + a)q(Y + a)$. We have $p(Y + a) = s(Y)$ and $q(Y + a) = t(Y)$ for some $s(Y), t(Y) \in R[Y]$. Thus $g(Y) = s(Y)t(Y)$. Since $p(X)$ and $q(X)$ are not units of $R[X]$, we know that $s(Y)$ and $t(Y)$ are not units of $R[Y]$. Therefore $g(Y) = s(Y)t(Y)$ is a proper factorization of $g(Y)$ in $R[Y]$. Similarly a proper factorization of $g(Y)$ gives rise to a proper factorization of $f(X)$.   ∎

Example 8.28.   Let $f(X) = X^7 + 7X - 1$. We shall show that $f(X)$ is irreducible over $\mathbb{Q}$. Note that we cannot apply the Eisenstein criterion to $f(X)$, and it would be rather tedious to consider the possible ways of writing $f(X)$ as a product of two polynomials of lower degree. In a situation like this it is worth trying to find a simple change of variable so that the Eisenstein criterion does apply to the new polynomial, and use 8.27.

Set $X = Y + 1$. Then $f(X) = g(Y)$ where

$$g(Y) = (Y + 1)^7 + 7(Y + 1) - 1$$

$$= Y^7 + 7Y^6 + 21Y^5 + 35Y^4 + 35Y^3 + 21Y^2 + 14Y + 7.$$

Thus $g(Y)$ is irreducible over $\mathbb{Q}$, by the Eisenstein criterion with $p = 7$. Therefore $f(X)$ is irreducible over $\mathbb{Q}$ by 8.27.   ■

We conclude the chapter by proving that two particular classes of polynomials are irreducible over $\mathbb{Q}$. These results will be needed in Chapter 12.

Theorem 8.29.   *Let p be a prime number. Set*

$$m(X) = X^{p-1} + x^{p-2} + \cdots + X^2 + X + 1.$$

Then $m(X)$ is irreducible over $\mathbb{Q}$.

Proof.   We cannot apply the Eisenstein criterion directly to $m(X)$, but we shall show that it can be applied after a suitable change of variable. Note that we have $m(X) = (X^p - 1)/(X - 1)$. Set $X = Y + 1$. Then

$$m(X) = ((Y + 1)^p - 1)/Y = (Y^p + pY^{p-1} + \cdots + pY + 1 - 1)/Y$$

$$= Y^{p-1} + pY^{p-2} + \frac{p(p-1)}{2} Y^{p-3} + \cdots + \frac{p(p-1)}{2} Y + p.$$

We shall denote the last polynomial in Y by $g(Y)$. The leading coefficient of $g(Y)$ is 1, and the other coefficients p, $p(p-1)/2,\ldots$ are binomial coefficients which are all integers divisible by p (see Question 7, Exercises 9). Also the constant term of $g(Y)$ is p which is not divisible by p^2. Therefore $g(Y)$ is irreducible over $\mathbb{Q}$ by the Eisenstein criterion. Hence $m(X)$ is irreducible over $\mathbb{Q}$ by 8.27.   ■

Theorem 8.30.   *Let p be a prime number. Set*

$$m(X) = X^{(p-1)p} + X^{(p-2)p} + \cdots + X^{2p} + X^p + 1.$$

Then $m(X)$ is irreducible over $\mathbb{Q}$.

Proof. Set $q = p^2$. We have

$$m(X) = (X^p)^{p-1} + (X^p)^{p-2} + \cdots + (X^p)^2 + X^p + 1$$
$$= ((X^p)^p - 1)/(X^p - 1).$$

We shall employ a change of variable as in the proof of 8.29, but this time the details are more complicated and we shall leave some of them to be filled in by the reader.

Let $m_p(X) = (X^p)^{p-1} + (X^p)^{p-2} + \cdots + (X^p)^2 + X^p + 1$ where the co-efficients are considered to be elements of $\mathbb{Z}/p\mathbb{Z}$. Thus $m(X)$ and $m_p(X)$ look the same but have coefficients in different rings. Set $X = Y + 1$. Then both $m(X)$ and $m_p(X)$ can be written as polynomials in Y which we shall call $g(Y)$ and $g_p(Y)$ respectively. In particular we have $g_p(Y) = ((Y+1)^q - 1)/((Y+1)^p - 1)$. However, when we work over $\mathbb{Z}/p\mathbb{Z}$ we have $(Y+1)^q = Y^q + 1$ and $(Y+1)^p = Y^p + 1$ (see Question 9, Exercises 9). Therefore $g_p(Y) = Y^q/Y^p = Y^{(p-1)p}$. This means that when we work over $\mathbb{Z}$ we have

$$g(Y) = Y^{(p-1)p} + a_{p-2}Y^{(p-2)p} + \cdots + a_2 Y^{2p} + a_1 Y^p + a_0$$

where the a_i are integers all of which are divisible by p. Also since $g(Y) = (Y+1)^{(p-1)p} + (Y+1)^{(p-2)p} + \cdots + (Y+1)^p + 1$, we have $a_0 = g(0) = p$. Therefore $g(Y)$ is irreducible over $\mathbb{Q}$ by the Eisenstein criterion, and hence $m(X)$ is also irreducible over $\mathbb{Q}$. ∎

Exercises 8

1. Working over $\mathbb{Z}$, find the content and the primitive of

$$(4/3)X^5 + (6/5)X^2 + 2.$$

2. Factorize each of the following polynomials into irreducible factors over the given ring. Give a justification in each case.
    (i) $X^4 + 4$ over $\mathbb{Z}$
    (ii) $X^4 + 4$ over $\mathbb{Z}/3\mathbb{Z}$
    (iii) $X^4 + 4$ over $\mathbb{Z}/13\mathbb{Z}$
    (iv) $X^2 + 3$ over $\mathbb{Z}/7\mathbb{Z}$
    (v) $X^2 + 1$ over $\mathbb{Z}/7\mathbb{Z}$
    (vi) $X^3 + 2$ over $\mathbb{Z}/3\mathbb{Z}$
    (vii) $X^5 - 10X + 2$ over $\mathbb{Q}$
    (viii) $X^5 - 10X + 1$ over $\mathbb{Q}$
    (ix) $6X^2 - 24$ over $\mathbb{Z}$
    (x) $X^3 + X^2 - 2X - 1$ over $\mathbb{Q}$
    (xi) $X^4 + 3X^3 - 3X - 2$ over $\mathbb{Z}/2\mathbb{Z}$
    (xii) $X^4 + 3X^3 + 3X - 2$ over $\mathbb{Z}$

3. Use Euler's criterion to determine whether $X^2 - 3$ is irreducible over $\mathbb{Z}/79\mathbb{Z}$.

4. Prove that $8X^3 - 6X - 1$ is irreducible over $\mathbb{Z}/7\mathbb{Z}$ and deduce that $8X^3 - 6X - 1$ is irreducible over $\mathbb{Z}$. (Compare this with Example 8.10.)

5. Find all the irreducible polynomials of degree 2 or 3 over $\mathbb{Z}/2\mathbb{Z}$.

6. Let R be an integral domain. Show that
   (i) the irreducible elements of R are irreducible as elements of $R[X]$.
   (ii) if $f(X) = g(X)h(X)$ in $R[X]$ with $f(X)$ primitive, then $g(X)$ and $h(X)$ are primitive.

9 Fields and field extensions

We have already given the definition of a field in 2.4, namely that it is a ring in which every non-zero element has a multiplicative inverse and $1 \neq 0$. We shall now go rather more deeply into the theory of fields. In many ways a field is the best sort of system to work with, because it is always possible to divide by a non-zero element. The material of this chapter has many uses in the remainder of the book. In particular it is crucial to the study of roots of polynomial equations. This is of great importance in such areas as number theory and geometry.

Examples 9.1. We have already met the following fields.

(a) The complex numbers $\mathbb{C}$.

(b) The real numbers $\mathbb{R}$.

(c) The rational numbers $\mathbb{Q}$.

(d) $\mathbb{Z}/p\mathbb{Z}$ when p is prime.

(e) The Gaussian numbers, i.e. all complex numbers $a + ib$ with $a, b \in \mathbb{Q}$.

(f) The quotient field of any integral domain, e.g. the field $F(X)$ of rational functions in X over a field F. ∎

Definition 9.2. Let F b a non-empty subset of a field K. Then F is said to be a *subfield* of K if F is a field with respect to the operations given in K. This situation is also described by such phrases as 'K is an *extension field* of the field F', or 'K is an *extension* of F'.

Examples 9.3

(a) $\mathbb{Q}$ is a subfield of $\mathbb{C}$.

(b) $\mathbb{R}$ is a subfield of $\mathbb{C}$.

(c) $\mathbb{Q}$ is a subfield of $\mathbb{R}$.

(d) Any field F is a subfield of the corresponding rational function field $F(X)$. ∎

Proposition 9.4.   *Let F be a non-empty subset of a field K. Then F is a subfield of K if and only if*

(i) *F contains the special elements 0 and 1 of K,*

(ii) *F is closed under subtraction and multiplication, and*

(iii) *every non-zero element of F has a multiplicative inverse in F.*

Proof.   Suppose that F satisfies (i), (ii) and (iii). Since the special elements 0 and 1 belong to F, they serve as 0 and 1 of F. In particular $0 \neq 1$ in F since this is the case in K. It follows from (i) and (ii) that F is a subring of K and hence is a ring in its own right. Also every non-zero element of F has an inverse in F by (iii). Therefore F is a subfield of K.

Conversely, suppose that F is a subfield of K. Question 3, Exercises 9 shows that F and K have the same 0 and 1. Clearly F satisfies (i), (ii) and (iii).   ∎

Comment 9.5.   Let a be a non-zero element of a field K. Since K is a field, a has an inverse a^{-1} in K by definition. Hence condition (iii) in 9.4 can be replaced by the requirement that for each non-zero $a \in F$, the element $a^{-1} \in F$.

Proposition 9.6.   *Every subfield of $\mathbb{C}$ contains $\mathbb{Q}$.*

Proof.   Let F be a subfield of $\mathbb{C}$. Then F contains the complex numbers 0 and 1 by 9.4, i.e. F contains the integers 0 and 1. Since F is closed under addition we have $2 = 1 + 1 \in F$. Hence $3 = 2 + 1 \in F$. Continuing this way we see that F contains every positive integer. Since F is also closed under taking negatives it follows that F contains every negative integer. Thus F contains $\mathbb{Z}$. Hence F contains a/b for all $a, b \in \mathbb{Z}$ with $b \neq 0$.   ∎

Definition 9.7.   Let F be a field. Then F is said to have positive *characteristic p* if p is the smallest positive integer such that

$$0 = 1 + 1 + \cdots + 1 \quad (p \text{ terms}).$$

If there is no such positive integer p, then F is said to have characteristic 0.

We shall soon see that the characteristic of a field plays an important role in its behaviour.

Example 9.8.   Let p be a prime number. The field $\mathbb{Z}/p\mathbb{Z}$ has characteristic p because 1 added to itself p times becomes 0 and p is clearly the smallest such integer.   ∎

Example 9.9. The complex number 1 added to itself any number of times never becomes zero. Therefore all subfields of $\mathbb{C}$ have characteristic 0. ∎

Notation. We shall denote the multiplicative identity of a field F by 1_F whenever there is a possibility of confusion with the integer 1. Thus $n1_F$ will denote 1_F added to itself n times. So for example, $3 \cdot 1_F = 1_F + 1_F + 1_F$.

Proposition 9.10. *Let F be a field of positive characteristic p. Then p is a prime number.*

Proof. We have $p \neq 1$ because $1_F \neq 0$. Let a, b be positive integers smaller than p. Recall that p is minimal with respect to the property that $p1_F = 0$. Hence $a1_F \neq 0 \neq b1_F$ as $a < p$ and $b < p$. Since F is an integral domain (Exercises 9, Question 2), we have $0 \neq (a1_F)(b1_F)$. However $(a1_F)(b1_F)$ is the product of a lots of 1_F by b lots of 1_F. Hence $(a1_F)(b1_F) = ab1_F$ where $ab1_F$ denotes the sum of ab lots of 1_F. Thus $ab1_F \neq 0$ so that $ab \neq p$. Therefore p is prime. ∎

Proposition 9.11. *Let F be a finite field. Then F has characteristic p for some prime p.*

Proof. Since F is a finite set the elements $1_F, 2 \cdot 1_F, \ldots, n1_F, \ldots$ cannot all be different. Hence there are distinct positive integers a and b such that $a1_F = b1_F$. Without loss of generality we may suppose that $a < b$. We have $0 = b1_F - a1_F = (b - a)1_F$. Thus there is a positive integer n with $n1_F = 0$, namely $n = b - a$. Therefore there is a smallest positive integer p such that $p1_F = 0$ and p is a prime number by 9.10. ∎

Proposition 9.12. *Let F be a field of positive characteristic p and let $x \in F$. Then $px = 0$ where px denotes x added to itself p times.*

Proof. We have

$$px = x + x + \cdots + x = 1_F x + 1_F x + \cdots + 1_F x = (1_F + 1_F + \cdots + 1_F)x$$

where the sum in each step is of p terms. Thus $px = (p1_F)x = 0 \cdot x = 0$. ∎

Proposition 9.13. Let F be a field of positive characteristic p. Then F has a subfield which looks just like $\mathbb{Z}/p\mathbb{Z}$.

Proof. Let a and b be positive integers with $a < b < p$. Since $0 < b - a < p$ we have $(b - a)1_F \neq 0$. It follows that $a1_F \neq b1_F$. Thus the elements $0, 1_F, 2 \cdot 1_F, 3 \cdot 1_F \ldots, (p - 1)1_F$ form a set S consisting of p distinct elements of F. These elements behave just like the corresponding elements $0, 1, 2, \ldots, p - 1$ of $\mathbb{Z}/p\mathbb{Z}$. The fact that $p1_F = 0$ corresponds to the fact that $p = 0$ in $\mathbb{Z}/p\mathbb{Z}$. It follows as in 5.7 that S is a subfield of F which looks just like $\mathbb{Z}/p\mathbb{Z}$. ∎

The expression 'looks like' can be made precise. See the definition of 'isomorphism' (13.7). This comment also applies to the next result.

Proposition 9.14. Let F be a field of characteristic 0. Then F has a subfield which looks just like $\mathbb{Q}$.

Proof. Let a and b be integers with $a < b$. Then $(b - a)1_F \neq 0$ and so $a1_F \neq b1_F$. It follows that the elements

$$\ldots, -3 \cdot 1_F, -2 \cdot 1_F, -1_F, 0, 1_F, 2 \cdot 1_F, 3 \cdot 1_F, \ldots$$

are all distinct. It is easy to see that they form a subring R of F. The subring R looks just like $\mathbb{Z}$. Let K be the subset of F consisting of all $a1_F/b1_F$ for $a1_F, b1_F \in R$ with $b1_F \neq 0$. Then K is a subfield of F which looks just like $\mathbb{Q}$ where $a1_F/b1_F$ corresponds to a/b for all $a, b \in \mathbb{Z}$ with $b \neq 0$. ∎

Summary 9.15. All the following statements are proved somewhere in this book.

(a) A field of characteristic 0 is infinite and has a subfield isomorphic to $\mathbb{Q}$ (9.14).

(b) If a field has positive characteristic p then p is a prime number (9.10).

(c) Every finite field has positive characteristic (9.11).

(d) An infinite field can have positive characteristic (9.16).

(e) If F is a finite field then F has p^n elements for some prime p and positive integer n (9.23).

(f) For each prime p and positive integer n there is one and only one field with p^n elements (15.9 and 15.10).

Example 9.16. Let p be a prime number. Set $F = \mathbb{Z}/p\mathbb{Z}$ and let $K = F(X)$ be the field of rational functions (see 7.4) in X over F. Then K is an infinite field of characteristic p. ∎

We now consider the properties of field extensions. Let K be an extension of the field F, i.e., K is a field and F is a subfield of K. We shall often want to regard K as being a vector space over the field F. In this context we shall call the elements of K 'vectors' even though they may not look anything like geometrical vectors. In order to make K into a vector space over F we need two basic ingredients: (i) We need a well-behaved way of adding vectors, i.e. of adding elements of K. We have this because K is a field and so has well-behaved addition. (ii) We need a well-behaved way of multiplying vectors by scalars, i.e. of multiplying elements of K by elements of F. We have this because there is a well-behaved way of multiplying any two elements of K. With these meanings for 'vector addition' and 'multiplication by scalars' we can check that K is a vector space over the field F. For example, let k be a vector and let a and b be scalars; i.e. let $k \in K$ and let $a, b \in F$. We require $(ab)k = a(bk)$, where $(ab)k$ means 'multiply the vector k by the scalar ab' and $a(bk)$ means 'first multiply the vector k by the scalar b, and then multiply the resulting vector by the scalar a'. However, the two are equal because the multiplication in K is associative.

Example 9.17. The field $\mathbb{C}$ of complex numbers is an extension of the field $\mathbb{R}$ of real numbers. In this case there is a convenient geometrical interpretation. The complex number $x + iy$ corresponds to the position vector (x, y) in the Argand diagram or real plane. Thus $\mathbb{C}$ can be thought of as a 2-dimensional vector space over $\mathbb{R}$. Clearly 1 and i form a basis for this vector space because every complex number is uniquely of the form $x \cdot 1 + y \cdot i$ with $x, y \in \mathbb{R}$. ∎

Example 9.18. The field $\mathbb{R}$ of real numbers is an extension of the field $\mathbb{Q}$ of rational numbers. As above we can regard $\mathbb{R}$ as being a vector space over $\mathbb{Q}$ and we just have to accept this in an abstract way beeause here there is no convenient geometrical interpretation. ∎

Example 9.19. Let K be the field of Gaussian numbers, i.e. all complex numbers $a + ib$ with $a, b \in \mathbb{Q}$. Then K is an extension of $\mathbb{Q}$. The corresponding vector space is 2-dimensional because 1 and i form a basis. ∎

We shall have many more examples later, particularly in Chapter 11.

Definition 9.20. (Dimension or degree of an extension) Let K be an extension of the field F. The *dimension* or *degree* of K over F is defined to be the dimension of K as a vector space over F. This dimension is denoted by $[K : F]$.

Comments 9.21. Let K be an extension of a field F.

(a) Since K contains a non-zero element (namely 1), it is not the zero vector space. Hence $[K:F] \neq 0$. Thus if $[K:F]$ is finite, it is a *positive* integer.

(b) K may be infinite dimensional (i.e. not finite dimensional) over F. Thus, for example, $\mathbb{R}$ is infinite dimensional over $\mathbb{Q}$.

(c) The word *degree* is used in this context because the dimension of this vector space often coincides with the degree of a certain polynomial (see 11.10).

(d) $[K:F] = 1$ if and only if $K = F$ (Question 11, Exercises 9).

Examples 9.22

(a) $[\mathbb{C}:\mathbb{R}] = 2$ by 9.17.

(b) If K is the field of Gaussian numbers, $[K:\mathbb{Q}] = 2$ by 9.19.

(c) Let t be a complex number such that $t^3 = 2$. Let K be the set of all complex numbers of the form $a + bt + ct^2$ with $a, b, c \in \mathbb{Q}$; it will be shown in 11.14 that K is a field, $[K:\mathbb{Q}] = 3$, and $1, t, t^2$ form a basis for K as a vector space over $\mathbb{Q}$. ∎

Theorem 9.23. *Let F be a finite field. Then F has p^n elements for some prime p and positive integer n.*

Proof. By 9.11 we know that F has characteristic p for some prime p. Hence by 9.13 F has a subfield S with p elements. As above we can regard F as being a vector space over the field S. Since F is a finite set, this vector space must have finite dimension. Thus $[F:S] = n$ for some positive integer n. Let $x_1, \ldots, x_n$ form a fixed basis for F as a vector space over S. Each element of F is *uniquely* of the form $s_1 x_1 + \cdots + s_n x_n$ for some $s_i \in S$. There are p possible values for each s_i, so that there are p^n possibilities for $s_1 x_1 + \cdots + s_n x_n$. ∎

The following theorem will be of fundamental importance in the text and of great use in problem solving. It applies to the 'three fields' context where there is a field K with a subfield H, and F is a subfield of H. Thus the fields increase in size in the order $F \subseteq H \subseteq K$. There are three field extensions which we can consider, namely K as an extension of F, K as an extension of H and H as an extension of F. The corresponding dimensions $[K:F]$, $[K:H]$ and $[H:F]$ are linked by the formula in (iii) below. We normally use the formula only when all the dimensions are finite.

Theorem 9.24. *Let F be a subfield of a field H which is a subfield of a field K.*

(i) *If $[K:F]$ is finite then $[K:H]$ and $[H:F]$ are finite.*

(ii) *If $[H:F]$ and $[K:H]$ are finite then $[K:F]$ is finite.*

(iii) *$[K:F]=[K:H][H:F]$ where the equation is interpreted to mean that if the left-hand side is infinite then so is at least one of the terms on the right-hand side and conversely.*

Proof. (i) Suppose that $[K:F]$ is finite. Since H is a subspace of K as a vector space over F, it follows that $[H:F]$ is also finite. Let $x_1,\ldots,x_n$ form a basis for K as a vector space over F. Every element of K is of the form $f_1x_1 + \cdots + f_nx_n$ for some $f_i \in F$. Clearly $f_i \in H$ so that $x_1,\ldots,x_n$ span, or generate, K as a vector space over H. Thus K has a finite spanning set as a vector space over H, and so $[K:H]$ is finite.

(ii) Suppose that $[K:H]$ and $[H:F]$ are finite. Set $s=[K:H]$ and $t=[H:F]$. Let $x_1,\ldots,x_s$ form a basis for K as a vector space over H, and let $y_1,\ldots,y_t$ form a basis for H as a vector space over F. We shall show that the st elements of the form x_iy_j form a basis for K as a vector space over F. This will show not only that $[K:F]$ is finite but also that $[K:F]=[K:H][H:F]$.

Suppose that

$$f_{11}x_1y_1 + f_{12}x_1y_2 + f_{21}x_2y_1 + \cdots + f_{st}x_sy_t = 0 \text{ for some } f_{ij} \in F. \quad (*)$$

We must show that $f_{ij}=0$ for all i and j. Set $h_i=f_{i1}y_1 + \cdots + f_{it}y_t$ for each $i=1,\ldots,s$. Then $h_i \in H$ since all f_{ij}, $y_j \in H$. We can rewrite $(*)$ in the form $h_1x_1 + \cdots + h_sx_s = 0$. Since the x_i are linearly independent over H we have $h_i=0$ for all i. Thus for each i we have $f_{i1}y_1 + \cdots + f_{it}y_t = 0$. Since the y_j are linearly independent over F we have $f_{ij}=0$ for all i and j. Thus the elements x_iy_j are linearly independent over F. Now let $k \in K$. As the x_i form a basis for K over H we have $k=a_1x_1 + \cdots + a_sx_s$ for some $a_i \in H$. Also the y_j form a basis for H over F. Hence for each i we have $a_i=b_{i1}y_1 + \cdots + b_{it}y_t$ for some $b_{ij} \in F$. Thus

$$k = a_1x_1 + \cdots + a_sx_s$$

$$= b_{11}x_1y_1 + b_{12}x_1y_2 + b_{21}x_2y_1 + \cdots + b_{st}x_sy_t.$$

Therefore the elements x_iy_j span K as a vector space over F.

(iii) This follows immediately from (i) and the way we proved (ii).   ∎

Corollary 9.25. *Let F be a subfield of a field K and suppose that $[K:F]$ is a prime number. Then the only subfields of K which contain F are K and F.*

Proof. Set $p = [K:F]$ and let H be a subfield of K and $H \supseteq F$. By 9.24 we have $p = [K:H][H:F]$. Since p is a prime number we have either $[K:H] = 1$ in which case $H = K$ or $[H:F] = 1$ in which case $H = F$ (see Question 11, Exercises 9). ∎

Definition 9.26. 'K is a *finite extension* of F' means that the field F is a subfield of the field K and the dimension $[K:F]$ of K as a vector space over F is finite. It does *not* mean that K is a finite set.

Exercises

1. Show that every finite-dimensional vector space over a countable field is countable. Hence, or otherwise, show that $\mathbb{R}$ is not finite dimensional as an extension of $\mathbb{Q}$.

2. Show that every field is an integral domain.

3. Let K be a field with a subfield F. Show that F and K have the same 0 and 1. (*Hint:* If H is a subgroup of a group G then H and G have the same identity element.)

4. Let F be a finite field with q elements and let F^* be the multiplicative group of non-zero elements of F. Prove that $x^{q-1} = 1$ for all $x \in F^*$ and hence $x^q = x$ for all $x \in F$. (*Hint:* If G is a finite group of order n then the order of each element of G divides n.)

5. Let F be a field. Prove that the rational function field $F(X)$ is an infinite-dimensional extension of F.

6. Let F be a field and S a non-empty subset of F. Let K be the intersection of all subfields of F which contain S. Prove that K is a subfield of F which contains S, and that if L is a subfield of F which contains S then L contains K. (K is called the *smallest subfield* of F containing S or the *subfield* of F *generated by S*.)

7. Let p be a prime number and r an integer in the range from 1 to $p - 1$. Prove that the binomial coefficient $p!/r!(p-r)!$ is divisible by p.

8. Let F be a field of positive characteristic p and let $a, b \in F$. Use the previous question to show that $(a + b)^p = a^p + b^p$. Let n be a positive integer and set $q = p^n$. Show that $(a + b)^q = a^q + b^q$.

9. Let F be a finite field with q elements and let $f(X)$ be a polynomial with coefficients in F. Show that $(f(X))^q = f(X^q)$. (*Hint:* Use questions 4 and 8.)

10. Let K be a finite field of *order p^n* (i.e. a field with p^n elements). Let F be a subfield of K of order p^r. By considering K as a vector space over F, or otherwise, show that $r \mid n$. Prove also that F is the only subfield of K of order p^r. (*Hint:* The elements of F are roots in K of the equation $X^q = X$ where $q = p^r$.)

11. Let K be an extension of a field F. Prove that $[K:F] = 1$ if and only if $K = F$.

12. Let K and L be finite extensions of a field F and suppose that the positive integers $[K:F]$ and $[L:F]$ are relatively prime. Let $H = K \cap L$. Prove that $H = F$.

10 Finite cyclic groups and finite fields

The main result of this chapter is that if F is a field and G is a finite subgroup of the multiplicative group of non-zero elements of F then G is cyclic. The proof is an elegant combination of elementary group theory and the theory of polynomial equations over a field. An immediate consequence of the main theorem is that if p is a prime number then the group of non-zero elements of $\mathbb{Z}/p\mathbb{Z}$ under multiplication is cyclic. We shall also prove some general results about finite cyclic groups. For instance, if G is a cyclic group with n elements then G has precisely one subgroup of order d for each positive divisor d of n.

Summary 10.1. We start by recalling some basic material from group theory. Let G be a multiplicatively written group with identity element 1. An element $x \in G$ is said to have *finite order* if $x^k = 1$ for some positive integer k. Suppose that x has finite order. Let n be the smallest positive integer such that $x^n = 1$. We say that x has *order* n and write $o(x) = n$. If x has finite order n and k is an integer, then $x^k = 1$ if and only if $n \mid k$. We have $o(x) = 1$ if and only if $x = 1$. If x has finite order n then the subgroup of G generated by x consists of the distinct elements $1, x, x^2, \ldots, x^{n-1}$. To say that G is *Abelian* means that $ab = ba$ for all $a, b \in G$. If G is Abelian then $(ab)^k = a^k b^k$ for every integer k and elements a and b of G. In fact $(ab)^k = a^k b^k$ if $ab = ba$ even if G is not Abelian. The group G is said to be *cyclic* if there is an element $x \in G$ such that every element of G is a power of x. In these circumstances we say that x *generates* G or that x is a *generator* of G. An infinite cyclic group is isomorphic to the group of integers under addition. A finite cyclic group of order n is isomorphic to the group of integers modulo n under addition.

The operation in a group will be assumed to be multiplication unless stated to the contrary. The number of elements in a set S is called the *order* of S and will be denoted by $|S|$. The order of G is a particularly important concept when G is a group.

Proposition 10.2. *Let G be a group with elements x and y of finite orders a and b respectively. Suppose that $xy = yx$ and that $(a, b) = 1$ (i.e. that a and b are relatively prime). Then xy has order ab.*

Proof. We have $(xy)^{ab} = x^{ab}y^{ab} = (x^a)^b(y^b)^a = 1^b \cdot 1^a = 1$. Thus xy has finite order. Set $n = o(xy)$. Since $(xy)^{ab} = 1$ we know that $n \mid ab$. However, $1 = 1^b = ((xy)^n)^b = x^{nb}y^{nb} = x^{nb}$ because $y^b = 1$. Thus $x^{nb} = 1$. Therefore $o(x) = a \mid nb$. Since $(a, b) = 1$, it follows that $a \mid n$. Similarly $b \mid n$. As a and b are relatively prime, and both a and b divide n, it follows that $ab \mid n$. Thus n and ab are positive integers, each of which divides the other. Therefore $n = ab$.   ∎

Proposition 10.3. *Let G be a group and let $x \in G$ be an element of finite order n. Suppose that $n = ab$ where a and b are relatively prime positive integers. Then there are elements y and $z \in G$ such that $o(y) = a$, $o(z) = b$ and $x = yz$. In fact we can take y and z to be powers of x.*

Proof. By 3.5 there are integers r and s such that $1 = ra + sb$. Set $y = x^{sb}$ and $z = x^{ra}$. We have $yz = x^{sb+ra} = x$. Also $y^a = x^{sba} = x^{sn} = (x^n)^s = 1^s = 1$. Hence y has finite order, say $o(y) = k$. Then $k \mid a$. But $1 = y^k = x^{sbk}$ and $o(x) = ab$. Hence $ab \mid sbk$ and so $a \mid sk$. However, $k = (ra + sb)k = rak + sbk$. Also $a \mid rak$ and $a \mid sbk$. So $a \mid k$ and hence $a = k$. Thus $o(y) = a$, and similarly $o(z) = b$.   ∎

Example 10.4. The point of this example is to illustrate with particular numbers the general method of proof which will be given in 10.5. Let G be an Abelian group of order 600. We shall show that G has an element x such that $o(y) \mid o(x)$ for all $y \in G$. We have $600 = 2^3 \cdot 3 \cdot 5^2$. Each element of G has finite order which divides 600, by Lagrange's theorem. There is at least one element of G whose order is a power of 2, namely the identity element which has order 2^0. Let $a \in G$ be such that $o(a)$ is a power of 2. Then $o(a) = 1, 2, 3$ or 8. Choose a so that $o(a)$ is as large as possible. The actual value of $o(a)$ depends on which Abelian group of order 600 we have. Suppose though that $o(a) = 4$. Suppose also that G has an element b with $o(b) = 3$. Clearly 3 is the largest power of 3 which can occur as the order of an element of G. Suppose that G has an element c with $o(c) = 5$ and that G has no elements of order 25. Set $x = abc$. By repeated use of 10.2 we have $o(x) = o(a) \cdot o(b) \cdot o(c) = 60$. Now let y be an arbitrary element of G. We have $o(y) = 2^i 3^j 5^k$ where $i = 0, 1, 2$ or 3; $j = 0$ or 1; $k = 0, 1$ or 2. By repeated use of 10.3 we have $y = uvw$ for some $u, v, w \in G$ with $o(u) = 2^i$, $o(v) = 3^j$, $o(w) = 5^k$. However 4 is the largest power of 2 which occurs as the order of an element of G. Hence $2^i = 1, 2$ or 4 so that $2^i \mid 4$. Similarly $3^j \mid 3$ and $5^k \mid 5$. Therefore $2^i 3^j 5^k \mid 4 \cdot 3 \cdot 5$, i.e., $o(y) \mid 60$. Thus G has an element x such that $o(y) \mid o(x)$ for all $y \in G$.   ∎

Theorem 10.5. *Let G be a finite Abelian group. Then there exists an element $x \in G$ such that $o(y) \mid o(x)$ for all $y \in G$.*

Proof. We generalize the method of 10.4 omitting some of the detail. We have $|G| = p_1^{c_1} p_2^{c_2} \cdots p_r^{c_r}$ where the p_i are distinct primes and the c_i are positive integers. For each i choose $a_i \in G$ such that the order of a_i is maximal with respect to being a power of p_i. Set $x = a_1 a_2 \cdots a_r$. Let $y \in G$. We have $o(y) = p_1^{d_1} p_2^{d_2} \cdots p_r^{d_r}$ for some non-negative integers d_i. By 10.3 we have $y = b_1 b_2 \cdots b_r$ for some $b_i \in G$ with $o(b_i) = p_i^{d_i}$. Because of the way a_i was chosen we know that $p_i^{d_i} \mid o(a_i)$. However, by 10.2 we have $o(y) = o(b_1) \cdot o(b_2) \cdots \cdot o(b_r)$ and $o(x) = o(a_1) \cdot o(a_2) \cdots \cdot o(a_r)$. Therefore $o(y) \mid o(x)$. ∎

Theorem 10.6. *Let F be a field and F^* the group of non-zero elements of F under multiplication. Let G be a finite subgroup of F^*. Then G is cyclic.*

Proof. Since G is a finite Abelian group, by 10.5 we know that there is an element $x \in G$ such that $o(y) \mid o(x)$ for all $y \in G$. Set $k = o(x)$ and $n = |G|$. Let $y \in G$. Since $o(y) \mid k$ we have $y^k = 1$. Thus all the n elements of G are roots of the polynomial equation $X^k = 1$. As this equation has coefficients in the field F, by 5.11 the number of roots of the equation in F is at most $\partial(X^k - 1)$. Hence $n \le k$. However, $k \mid n$ by Lagrange's theorem. Therefore $k = n$. Thus $o(x) = |G|$ and so x generates G. ∎

Corollary 10.7. *Let F be a finite field. Then the group F^* of non-zero elements of F under multiplication is cyclic.*

Proof. Take $G = F^*$ in 10.6. ∎

Corollary 10.8. *Let p be a prime number. Then the group of non-zero elements of $\mathbb{Z}/p\mathbb{Z}$ under multiplication is cyclic of order $p - 1$.*

Proof. Note that $\mathbb{Z}/p\mathbb{Z}$ is a field by 5.7. ∎

Comments 10.9

(i) Let p be a prime number and let G be the group of non-zero elements of $\mathbb{Z}/p\mathbb{Z}$ under multiplication. We know by 10.8 that G is cyclic. In number theory, a generator for G is called a *primitive root modulo p*. For small values of p it is possible to find a generator for G by trial and error. There is usually more than one choice of generator. In general there is no formula or procedure known for finding a generator for G from the value of p.

(ii) Let n be a positive integer and let U be the group of units of the ring $\mathbb{Z}/n\mathbb{Z}$. Recall from 5.4 that the elements of U are those elements of $\mathbb{Z}/n\mathbb{Z}$ which, as integers, are relatively prime to n. Clearly U is a finite Abelian group. We know that if n is prime then U is cyclic, because then U is the multiplicative group of non-zero elements of the field $\mathbb{Z}/n\mathbb{Z}$. It is a standard result in number theory, which we shall not prove, that U is cyclic if and only if $n = 2$ or 4 or p^k or $2p^k$ for some odd prime p and positive integer k.

Example 10.10. Let G be the group of non-zero elements of $\mathbb{Z}/23\mathbb{Z}$ under multiplication. We know by 10.8 that G is a cyclic group of order 22. We shall find a generator for G. Thus we need an element $g \in G$ with $o(g) = 22$. Let $x \in G$. By Lagrange's theorem $o(x) | |G|$. Thus $o(x) = 1, 2,$ 11 or 22. Therefore finding a generator for G is equivalent to finding an element $x \in G$ such that x does not have order 1, 2 or 11. Recall that $o(x) = 1$ if and only if $x = 1$.

In order to find a generator for G we start by trying $x = 2$; clearly neither 1 nor -1 is a generator. Working in $\mathbb{Z}/23\mathbb{Z}$ we have $2 \neq 1$ and $2^2 = 4, 2^5 = 32 = 9, 2^{10} = 9^2 = 81 = 12 + 69 = 12, 2^{11} = 2 \cdot 12 = 24 = 1$. Thus 2 has order 11 as an element of G, so that 2 is not a generator of G. However, it is clear from the above calculations that $(-2)^2 = 4 \neq 1$ and $(-2)^{11} = -2^{11} = -1$. Thus -2 does not have order 1, 2 or 11, so that $o(-2) = 22$ and so -2 is a generator for G. In other words, every element of G is a power of -2. If it had turned out that neither 2 nor -2 was a generator for G, then we would next have tried 3 and -3 and so on. ∎

We shall next use 10.6 to prove the following important fact about subfields of finite fields.

Theorem 10.11. *Let F be a finite field with p^n elements for some prime p and positive integer n, and let r be a positive integer such that $r | n$. Then F has one and only one subfield of order p^r.*

Proof. Set $q = p^n$ and $t = p^r$. We must show that F has a subfield S with t elements. The uniqueness of S will follow from Question 10, Exercises 9. By 10.7 there is an element $x \in F$ such that x generates the multiplicative group F^* of non-zero elements of f. Thus $x^{q-1} = 1$ and the non-zero elements of F are $1, x, x^2, \ldots, x^{q-2}$. We have $n = rs$ for some positive integer s. Also

$$q - 1 = p^n - 1 = (p^r)^s - 1 = (p^r - 1)\left((p^r)^{s-1} + (p^r)^{s-2} + \cdots + p^r + 1\right).$$

Thus $q - 1 = (t - 1)k$ for some positive integer k.

Set $y = x^k$. We have $y^{t-1} = x^{(t-1)k} = x^{q-1} = 1$. Thus y has finite order as an element of the group F^* and $o(y)|t-1$. Set $a = o(y)$. Then $a|t-1$. We have $1 = y^a = x^{ka}$. Since $o(x) = q-1$ it follows that $q-1|ka$. Thus $(t-1)k|ka$ and so $t-1|a$. Therefore $a = t-1$, i.e. $o(y) = t-1$. Let S consist of the t distinct elements $0, 1, y, y^2, \ldots, y^{t-2}$. Since $y^{t-1} = 1$, the non-zero elements of S are roots of the equation $X^{t-1} = 1$. Hence every element of S is a root of the equation $X^t = X$. Since S contains t elements and $t = \partial(X^t - X)$, it follows by 5.11 that S consists precisely of those elements $f \in F$ such that $f^t = f$.

We can now forget how we constructed S and just use the facts that S has t elements and that S is the set of those elements $f \in F$ such that $f^t = f$. We shall complete the proof by showing that S is a subfield of F. Clearly 0 and $1 \in S$. Let $a, b \in S$. We have $a^t = a$ and $b^t = b$. Hence $(ab)^t = a^t b^t = ab$, so that $ab \in S$. Also if $a \neq 0$ we have $(1/a)^t = 1/a^t = 1/a$, so that $1/a \in S$. It remains to show that $a - b \in S$. If p is odd then t is odd and $(-b)^t = (-1)^t b^t = -b$. If $p = 2$ then $-f = f$ for all $f \in F$ so that $(-b)^t = b^t = b = -b$. Thus $(-b)^t = -b$ whatever the value of the prime p. Therefore by Question 8, Exercises 9 we have $(a-b)^t = (a+(-b))^t = a^t + (-b)^t = a - b$, so that $a - b \in S$. ∎

Summary 10.12. Let F be a finite field with p^n elements. Any subfield of F has p^r elements for some positive factor r of n, and for each such r there is precisely one subfield of F with p^r elements.

Lemma 10.13. *Let x be an element of finite order n in a group G and let i be an integer. Then $o(x^i) = n/(i, n)$ where (i, n) is the HCF of i and n.*

Proof. We have $(x^i)^n = (x^n)^i = 1^i = 1$. Hence x^i has finite order. Set $k = o(x^i)$ and $h = (i, n)$. We have $i = hi'$ and $n = hn'$ where $(i', n') = 1$. We must show that $k = n'$. We have $(x^i)^{n'} = x^{in'} = x^{hi'n'} = (x^n)^{i'} = 1^{i'} = 1$, so that $k|n'$. Also $1 = (x^i)^k = x^{ik}$. Hence $o(x)|ik$. Thus $n|ik$, i.e. $hn'|hi'k$ and so $n'|i'k$. However, $(i', n') = 1$. Therefore by 3.8 $n'|k$. Hence $n' = k$. ∎

Proposition 10.14. *Let x be a generator of a finite cyclic group G of order n and let i be a positive integer in the range from 1 to n. Then x^i is a generator of G if and only if $(i, n) = 1$.*

Proof. Using 10.13 it is easy to see the equivalence of the following statements:
(i) x^i generates G; (ii) $o(x^i) = n$; (iii) $n/(i, n) = n$; (iv) $(i, n) = 1$. ∎

Definition 10.15 (Euler's φ-function). The function φ is defined on the set of positive integers as follows: let n be a positive integer. Then $\varphi(n)$ is defined to be the number of positive integers in the range from 1 to n which are relatively prime to n.

Comments 10.16. Let n be a positive integer.

(a) We know by 10.14 that $\varphi(n)$ is the number of generators of a cyclic group of order n. Of course a cyclic group, by definition, has one generator; the phrase 'the number of generators of a cyclic group' means 'the number of elements of a cyclic group each of which is a generator of the group'.

(b) As in 10.9(ii) we know that $\varphi(n)$ is the number of units of the ring $\mathbb{Z}/n\mathbb{Z}$.

(c) In general $\varphi(ab) \neq \varphi(a)\varphi(b)$. However, $\varphi(ab) = \varphi(a)\varphi(b)$ if $(a, b) = 1$ (see 10.21).

(d) $\varphi(n) = n(1 - 1/p_1)(1 - 1/p_2) \cdots (1 - 1/p_r)$ where $p_1, p_2, \ldots, p_r$ are the distinct prime factors of n (see Question 11, Exercises 10).

(e) $\sum \varphi(d) = n$, where d ranges over the positive divisors of n (see 10.22). Thus for example,

$$\varphi(1) + \varphi(2) + \varphi(3) + \varphi(4) + \varphi(6) + \varphi(12) = 1 + 1 + 2 + 2 + 2 + 4 = 12.$$

Theorem 10.17. *Let G be a finite cyclic group of order n and let r be a positive factor of n. Then G has one and only one subgroup of order r.*

Proof. We have $n = rd$ for some positive integer d. Let x be a generator of G and set $y = x^d$. By 10.13 we have $o(y) = n/(d, n) = n/d = r$. Hence the subgroup Y of G generated by y has order r. Now let A be any subgroup of G of order r and let $a \in A$. By Lagrange's theorem $o(a) \,|\, |A|$. Thus $o(a) \,|\, r$ and so $a^r = 1$. However, $a \in G$ so that $a = x^i$ for some integer i. Thus $x^{ir} = 1$, i.e. $o(x) \,|\, ir$, i.e. $n \,|\, ir$. Hence $rd \,|\, ir$ and so $d \,|\, i$. Thus $i = jd$ for some integer j. We have $a = x^i = (x^d)^j = y^j$ so that $a \in Y$. Hence $A \subseteq Y$. Since $|A| = |Y|$ it follows that $A = Y$. ∎

Comments 10.18. The statements of 10.11 and 10.17 have a formal similarity in the sense that '$r \,|\, n$ implies the existence of a unique sub-object corresponding to r'. In one case it is a subfield of order p^r in the field of order p^n, in the other it is a subgroup of order r in a cyclic group of order n. Finite fields and finite groups can be linked by the fact that corresponding to a finite field of order p^n there is a cyclic group of order n, namely its automorphism group. The Frobenius automorphism defined in Chapter 13 is a generator of this group.

Notation 10.19. Let A and B be additively written Abelian groups. Then $A \oplus B$ denotes the group consisting of all pairs (a, b) for $a \in A$, $b \in B$ with addition of pairs given by $(a, b) + (a', b') = (a + a', b + b')$. The group $A \oplus B$ is called the *direct sum* (sometimes the *direct product* or the *Cartesian product*) of A and B.

Proposition 10.20. *Let a and b be relatively prime positive integers. Then the additive group $\mathbb{Z}/a\mathbb{Z} \oplus \mathbb{Z}/b\mathbb{Z}$ is cyclic and is isomorphic to $\mathbb{Z}/ab\mathbb{Z}$.*

Proof. Set $G = \mathbb{Z}/a\mathbb{Z} \oplus \mathbb{Z}/b\mathbb{Z}$. In what follows, the group operation is always addition. Let x and y be generators of the cyclic groups $\mathbb{Z}/a\mathbb{Z}$ and $\mathbb{Z}/b\mathbb{Z}$ respectively. We have $o(x) = |\mathbb{Z}/a\mathbb{Z}| = a$, and $o(y) = b$. Working in G we have $(x, y) = (x, 0) + (0, y)$ where $(x, 0)$ and $(0, y)$ have orders a and b respectively. By 10.2 we know that (x, y) has order ab as an element of G. Hence (x, y) generates G as $|G| = ab$. Thus G and $\mathbb{Z}/ab\mathbb{Z}$ are both cyclic groups of order ab and so they are isomorphic. ∎

Proposition 10.21. *Let a and b be relatively prime positive integers. Then $\varphi(ab) = \varphi(a)\varphi(b)$.*

Proof. Set $G = \mathbb{Z}/a\mathbb{Z} \oplus \mathbb{Z}/b\mathbb{Z}$. In what follows, the group operation is always addition. By 10.14 and 10.20 we know that $\varphi(ab)$ is the number of generators of the cyclic group $\mathbb{Z}/ab\mathbb{Z}$ and hence also of the cyclic group G. Thus it is enough to show that G has $\varphi(a)\varphi(b)$ generators. As was shown in the proof of 10.20, if x and y are generators of $\mathbb{Z}/a\mathbb{Z}$ and $\mathbb{Z}/b\mathbb{Z}$ respectively, then (x, y) is a generator of G. Now suppose that (u, v) is a generator of G for some $u \in \mathbb{Z}/a\mathbb{Z}$, $v \in \mathbb{Z}/b\mathbb{Z}$. Now let $w \in \mathbb{Z}/a\mathbb{Z}$. The element $(w, 0) \in G$ can be obtained by adding the generator (u, v) to itself a positive number of times, say i times. Thus $(w, 0) = (iu, iv)$ so that $w = iu$. Hence every element of $\mathbb{Z}/a\mathbb{Z}$ can be obtained by adding u to itself a finite number of times. Therefore u is a generator of $\mathbb{Z}/a\mathbb{Z}$ and similarly v is a generator of $\mathbb{Z}/b\mathbb{Z}$. Thus we have shown that the generators of G are precisely the elements of the form (u, v) where u is a generator of $\mathbb{Z}/a\mathbb{Z}$ and v is a generator of $\mathbb{Z}/b\mathbb{Z}$. However, $\mathbb{Z}/a\mathbb{Z}$ has $\varphi(a)$ generators and $\mathbb{Z}/b\mathbb{Z}$ has $\varphi(b)$ generators. Therefore G has $\varphi(a)\varphi(b)$ generators and so $\varphi(ab) = \varphi(a)\varphi(b)$. ∎

Proposition 10.22. *Let n be a positive integer. Then $n = \sum \varphi(d)$ where d ranges over the positive divisors of n.*

Proof. Let G be a cyclic group of order n. We partition the n elements of G by their orders. Thus for each positive divisor d of n let S_d be the set of elements of G of order d. Let G_d be the unique subgroup of G of order d and let $x \in S_d$. Since $o(x) = d$ we know that x generates a subgroup of G of order d. Hence x generates G_d. Conversely, every generator of G_d has order d and so is an element of S_d. Therefore the number of elements in S_d is the number of generators of the cyclic group G_d. Hence by 10.14 $|S_d| = \varphi(d)$. Therefore $n = \sum |S_d| = \sum \varphi(d)$.   ∎

Exercises 10

1. Let G be an Abelian group and let T be the set of elements of G which have finite order. Prove that T is a subgroup of G.

2. Let F^* be the multiplicative group of non-zero elements of a field F. Find the elements of finite order in F^* when f is (i) the complex numbers $\mathbb{C}$, (ii) the real numbers $\mathbb{R}$, (iii) the rational numbers $\mathbb{Q}$.

3. Give an example of a group G with elements x and y of finite order such that $o(xy) \neq o(x)o(y)$.

4. Show that the group of units of $\mathbb{Z}/8\mathbb{Z}$ is not cyclic. Hence or otherwise show that if k is a positive integer with $k \geq 3$ then the group of units of $\mathbb{Z}/2^k\mathbb{Z}$ is not cyclic.

5. For $n = 2, 3, 4, 5, 6, 7, 9, 10, 11, 13, 14, 17, 18, 19$ verify directly that the group U of units of $\mathbb{Z}/n\mathbb{Z}$ is cyclic and list all the generators of U.

6. Verify that the group of units of $\mathbb{Z}/n\mathbb{Z}$ is not cyclic if $n = 12, 15$ or 20.

7. Let a and b be positive integers such that $(a, b) \neq 1$ and let G be the additive group $\mathbb{Z}/a\mathbb{Z} \oplus \mathbb{Z}/b\mathbb{Z}$. Prove that G is not cyclic. (*Hint:* Prove that the order of every element of G divides $ab/(a,b)$.)

8. (Euler's generalization of Fermat's little theorem) Let n be a positive integer and let a be an integer such that $(n, a) = 1$. Show that $a^{\varphi(n)} \equiv 1$ (mod n). (Note that $\varphi(n) = n - 1$ if n is a prime.)

9. Give an example of two positive integers a and b such that

$$\varphi(ab) \neq \varphi(a)\varphi(b).$$

10. Let p be a prime number and let n be a positive integer. Show that $\varphi(p^n) = p^{n-1}(p - 1)$. (*Hint:* Count how many of the integers $1, 2, 3, \ldots, p^n$ are not relatively prime to p^n.)

11. Let n be a positive integer. Use 10.21 and Question 10 to show that $\varphi(n) = n(1 - 1/p_1)(1 - 1/p_2) \cdots (1 - 1/p_r)$ where $p_1, p_2, \ldots, p_r$ are the distinct prime factors of n.

12. Find a generator for the group of non-zero elements of $\mathbb{Z}/p\mathbb{Z}$ under multiplication when (i) $p = 29$, (ii) $p = 31$.

11 Algebraic numbers

Many important numbers, such as $\sqrt{2}$ and i, are solutions of polynomial equations. Such numbers are said to be algebraic, and we shall study their basic theory in this chapter. Algebraic numbers also occur frequently in the following context: We start with a polynomial $f(X)$ over $\mathbb{Q}$ and we wish to study the solutions of the corresponding equation $f(X) = 0$. These solutions are algebraic numbers.

Definition 11.1. Let K be a field with a subfield F. Then $t \in K$ is said to be *algebraic* over F if $f(t) = 0$ for some non-zero polynomial $f(X)$ with coefficients in F.

Comments 11.2

(a) We shall often say 'let t be algebraic over F'. It will be understood that t belongs to some field extension of F.

(b) We shall leave till Chapter 14 the more difficult problem: Starting from a field F and a polynomial $f(X)$ with coefficients in F, construct a field K which contains F and also contains a root of the equation $f(X) = 0$.

(c) Let $f(X)$ be a polynomial with coefficients in $\mathbb{C}$. The fundamental theorem of algebra guarantees that all the roots of the equation $f(X) = 0$ are in $\mathbb{C}$ and that the equation has the right number of solutions in $\mathbb{C}$. Thus, for example, if $f(X)$ has coefficients in $\mathbb{Q}$ and if t is a root of the equation $f(X) = 0$, then we can work inside the field $\mathbb{C}$ because $\mathbb{C}$ contains both t and $\mathbb{Q}$.

(d) In number theory, an *algebraic number* is a complex number which is algebraic over $\mathbb{Q}$. Complex numbers which are not algebraic, such as e and π, are called *transcendental*.

(e) When talking about algebraic elements we must say over which field they are algebraic, or at least it should be clear from the context.

Examples 11.3

(a) $\sqrt{2}$ is algebraic over $\mathbb{Q}$ because it is a root of the equation $X^2 - 2 = 0$ which has coefficients in $\mathbb{Q}$.

(b) i is algebraic over $\mathbb{R}$ because i is a root of the polynomial equation $X^2 + 1 = 0$ over $\mathbb{R}$. This also shows that i is algebraic over $\mathbb{Q}$ because the coefficients also belong to $\mathbb{Q}$.

(c) Set $t = i + \sqrt{2}$. Then $t - i = \sqrt{2}$ so that $2 = (t - i)^2 = t^2 - 2it - 1$. Hence $2it = t^2 - 3$ so that $-4t^2 = (t^2 - 3)^2 = t^4 - 6t^2 + 9$. Therefore $i + \sqrt{2}$ is algebraic over $\mathbb{Q}$, being a root of the equation $X^4 - 2X^2 + 9 = 0$. ∎

Definitions 11.4

(i) A polynomial is said to be *monic* if its leading coefficient is 1.

(ii) Let F be a field and t an element of some extension of F. Then a polynomial $m(X)$ with coefficients in F is said to be the *minimum polynomial* of t over F if $m(X)$ is a monic polynomial of the lowest degree with its coefficients in F such that $m(t) = 0$.

(iii) To say that t is *algebraic of degree n* over F means that t is algebraic over F and its minimum polynomial has degree n.

Comments 11.5

(a) In 11.6 we shall show that the minimum polynomial of an algebraic element is unique, thus justifying the term 'the minimum polynomial'.

(b) An element t is algebraic of degree 1 over a field F if and only if $t \in F$ (see Question 1, Exercises 11).

(c) Since we are working over a field F, any non-zero polynomial with coefficients in F can be made monic by dividing out by its leading coefficient. For example, $2X^2 + 3$ can be made monic over $\mathbb{Q}$ by replacing it by $X^2 + 3/2$. Of course $X^2 + 3/2$ is not the same polynomial as $2X^2 + 3$ but the equations $2X^2 + 3 = 0$ and $X^2 + 3/2 = 0$ have the same roots. We could not make $2X^2 + 3$ monic if we had to keep the coefficients in $\mathbb{Z}$.

Theorem 11.6. *Let t be an element algebraic over a field F and with minimum polynomial $m(x)$. Then*

(i) *$m(X)$ is unique;*

(ii) *$m(X)$ is irreducible over F;*

(iii) *if $f(X)$ is a polynomial with coefficients in F and $f(t) = 0$ then $m(X) \mid f(X)$ (i.e. $f(X) = m(X)g(X)$ for some polynomial $g(X)$ with coefficients in F);*

(iv) *If $f(X)$ is a monic polynomial with coefficients in F such that $f(X)$ is irreducible over F and $f(t) = 0$, then $f(X) = m(X)$. (This gives a very useful method for proving that a particular polynomial $f(X)$ is the minimum polynomial of t.)*

Proof. (i) Suppose that $q(X)$ is also a monic polynomial with coefficients in F and that $q(X)$ has the smallest possible degree subject to $q(t) = 0$. Since both $m(X)$ and $q(X)$ have the smallest possible degree, we have $\partial(m) = \partial(q)$. Set $\partial(m) = n$ and let $g(X) = m(X) - q(X)$. Both $m(X)$ and $q(X)$ have leading term X^n. Hence either $\partial(g) < n$ or $g(X)$ is the zero polynomial. Clearly $g(X)$ has coefficients in F and $g(t) = m(t) - q(t) = 0$. Therefore $g(X)$ must be the zero polynomial, for otherwise we could obtain from $g(X)$ a monic polynomial of degrees less than n with t as a root. Thus $m(X) - q(X) = 0$ and so $m(X) = q(X)$, proving the uniqueness of $m(X)$.

(ii) Suppose that $m(X)$ is not irreducible over F. Then $m(X) = f(X)g(X)$, where $f(X)$ and $g(X)$ are polynomials with coefficients in F with $\partial(f) < \partial(m)$ and $\partial(g) < \partial(m)$. We have $0 = m(t) = f(t)g(t)$ so that $f(t) = 0$ or $g(t) = 0$. (Here we need to know that $f(t)$ and $g(t)$ are elements of some larger field.) This is a contradiction since $\partial(f) < \partial(m)$ and $\partial(g) < \partial(m)$.

(iii) Let $f(X)$ be a polynomial with coefficients in F and suppose that $f(t) = 0$. By the division algorithm for polynomials (2.9) we have $f(X) = q(X)m(X) + r(X)$ for some polynomials $q(X)$ and $r(X)$ with coefficients in F, where either $r(X)$ is the zero polynomial or $\partial(r) < \partial(m)$. We have $0 = f(t) = q(t)m(t) + r(t) = r(t)$. If $r(X)$ were not the zero polynomial we would have a contraction to the minimality of $\partial(m)$. Therefore $f(X) = q(X)m(X)$.

(iv) Let $f(X)$ be a monic irreducible polynomial with coefficients in F. Suppose that $f(t) = 0$. By (iii) we have $f(X) = m(X)g(X)$ for some polynomial $g(X)$ with coefficients in F. However, $f(X)$ is irreducible and $m(X)$ is not a constant. Therefore $g(X)$ must be a constant. Since $f(X)$ and $m(X)$ both have leading coefficient 1, we have $g(X) = 1$. ∎

Example 11.7. Suppose that an element t is algebraic over a field F. In order to find the minimum polynomial for t over F, we first find some monic polynomial $f(X)$ with coefficients in F such that $f(t) = 0$. We then test whether $f(X)$ is irreducible over F (see Chapter 8). If $f(X)$ is irreducible over F then $f(X)$ is the minimum polynomial for t over F by 11.6(iv). On the other hand suppose that $f(X)$ factorizes properly over F as $f(X) = g(X)h(X)$ where $\partial(g) < \partial(f)$ and $\partial(h) < \partial(f)$. Then $g(t) = 0$ or $h(t) = 0$ because $g(t)h(t) = f(t) = 0$. Without loss of generality we may suppose that $g(t) = 0$. We then repeat the procedure with $g(X)$ in place of $f(X)$ noting that $\partial(g) < \partial(f)$. This procedure will produce the minimum polynomial in a finite number of steps. In particular, take $F = \mathbb{Q}$ and $t = i + \sqrt{2}$. We know by 11.3(c) that $f(t) = 0$ where $f(X) = X^4 - 2X^2 + 9$. Also $f(X)$ is irreducible over $\mathbb{Q}$ by 8.11 and 8.22. Therefore $f(X)$ is the minimum polynomial for t over $\mathbb{Q}$, and t is algebraic of degree 4 over $\mathbb{Q}$ because $\partial(f) = 4$. $\blacksquare$

Definition 11.8. Let K be an extension of the field F and let $t \in K$. The intersection of all those subfields of K which contain both F and t is denoted by $F(t)$.

Thus $F(t)$ is a subfield of K. It is the 'smallest' subfield of K containing both F and t in the sense that if L is a subfield of K with $F \subseteq L$ and $t \in L$ then $F(t) \subseteq L$.

We say that $F(t)$ is F with t *adjoined*, or that we form $F(t)$ by *adjoining* t to F.

Lemma 11.9. *Let K be an extension of the field F and $t \in K$. Suppose that t is algebraic over F. Then $F(t)$ consists of all elements of K which can be written in the form $a_0 + a_1 t + \cdots + a_r t^r$ for some non-negative integer r and some $a_i \in F$.*

Proof. Let $m(X)$ be the minimum polynomial for t over F and let H be the subset $\{a_0 + a_1 t + \cdots + a_r t^r \mid a_i \in F, \ r \geq 0\}$ of K. Let $h \in H$. Then $h = a_0 + a_1 t + \cdots + a_r t^r$ for some $a_i \in F$ and $r \geq 0$. Since $F(t)$ is a field, $a_i \in F(t)$ and $t \in F(t)$ we have $h \in F(t)$ and so $H \subseteq F(t)$. Now clearly $F \subseteq H$ and $t \in H$. So in order to prove that $F(t) \subseteq H$, it is enough to show that H is a subfield of K.

We need only to show that the non-zero elements of H have inverses in H for the other conditions needed for it be a subfield of K are easily seen to hold. Let $h \in H$, $h \neq 0$. We have $h = a_0 + a_1 t + \cdots + a_r t^r$ for some $a_i \in F$ and an integer $r \geq 0$. Set $f(X) = a_0 + a_1 X + \cdots + a_r X^r$. Then $f(X)$ is a polynomial with coefficients in F and $f(t) = h \neq 0$. Suppose that $f(X) = m(X)g(X)$ for some polynomial $g(X)$ with coefficients in F. Since $m(t) = 0$ we then have $f(t) = m(t)g(t) = 0$, which is a contradiction. Thus, working in the ring S of polynomials in X over F, we know that $m(X)$ is irreducible (11.6(ii)) and as above does not divide $f(X)$. Hence $m(X)$ and $f(X)$ have no non-trivial common factors. Since S is Euclidean (2.9), there are polynomials $p(X)$ and $q(X)$ with coefficients in F such that $1 = p(X)m(X) + q(X)f(X)$ (3.5). Hence $1 = p(t)m(t) + q(t)f(t) = q(t)f(t)$. However, $q(t) = b_0 + b_1 t + \cdots + b_s t^s$ for some $b_i \in F$ and an integer $s \geq 0$. Therefore $q(t)$ is the multiplicative inverse of $f(t)$ and $q(t) \in H$. $\blacksquare$

Theorem 11.10. *Let K be an extension of a field F and let $t \in K$. Suppose that t is algebraic of degree n over F. Then*

(i) *$[F(t):F] = n$ where $[F(t):F]$ denotes as usual the degree or dimension of $F(t)$ as a vector space over F;*

(ii) *the elements of $F(t)$ can be expressed uniquely in the form*

$$a_0 + a_1 t + \cdots + a_{n-1} t^{n-1}$$

for some $a_i \in F$;

(iii) *the elements $1, t, t^2, \ldots, t^{n-1}$ form a basis for $F(t)$ as a vector space over F.*

Proof. We shall prove all three parts of the statement together. Let $m(X)$ be the minimum polynomial of t over F. Then $\partial(m) = n$. Let $w \in F(t)$. By 11.9 we know that $w = b_0 + b_1 t + \cdots + b_r t^r$ for some $b_i \in F$ and an integer $r \geq 0$. Set $f(X) = b_0 + b_1 X + \cdots + b_r X^r$. By the division algorithm for polynomials over F (2.9) we have $f(X) = q(X)m(X) + r(X)$ for some polynomials $q(X)$, $r(X)$ with coefficients in F, and where either $r(X)$ is the zero polynomial or $\partial(r) < \partial(m) = n$. We have

$$w = b_0 + b_1 t + \cdots + b_r t^r = f(t) = q(t)m(t) + r(t) = r(t)$$

because $m(t) = 0$. We can write $r(X) = a_0 + a_1 X + \cdots + a_{n-1} X^{n-1}$ for some $a_i \in F$. Here we might have $a_i = 0$ for all i, which corresponds to $r(X)$ being the zero polynomial. Also we might, for example, have $a_{n-1} = 0$ and $a_{n-2} \neq 0$ if $\partial(r) = n - 2$. Thus we have $w = a_0 + a_1 t + \cdots + a_{n-1} t^{n-1}$ for some $a_i \in F$.

It remains to show that this representation for w is unique. Suppose that we also have $w = c_0 + c_1 t + \cdots + c_{n-1} t^{n-1}$ for some $c_i \in F$. Set $g(X) = c_0 + c_1 X + \cdots + c_{n-1} X^{n-1}$. We have $r(t) = w = g(t)$ so that t is a root of the equation $r(X) - g(X) = 0$. Now either $r(X) - g(X)$ is the zero polynomial or it has degree at most $n - 1$. Since t is algebraic of degree n it follows that $r(X) - g(X)$ is the zero polynomial. Thus all the coefficients of $r(X) - g(X)$ are zero and so $c_i = a_i$ for all i. ∎

Comments 11.11. Let K be an extension of a field F and let $t \in K$ such that t is algebraic of degree n over F. Let $m(X)$ be the minimum polynomial of t over F.

(a) By 11.10 the dimension of $F(t)$ as a vector space over F coincides with the degree of the polynomial $m(X)$. This is why the term 'degree' is often used to mean 'dimension' in this context.

(b) Once we know $m(X)$ we can use 11.10 to give a concrete description of the elements of $F(t)$. An element of $F(t)$ can be written as a polynomial expression in t with coefficients in F, and this expression is not in general unique. For example if $t = \sqrt{2}$ then t and $t^3/2$ are both representations for $\sqrt{2}$ as polynomial expressions in t with coefficients in $\mathbb{Q}$. However, we can always arrange an element of $F(t)$ to be written as a polynomial expression in t using powers of t up to at most t^{n-1}, and this expression is unique. We do this by using the equation $m(t) = 0$ to express t^n and higher powers of t in terms of lower powers of t. For example, suppose that $m(X) = X^3 - 2X^2 + 3X + 4$. Then $t^3 - 2t^2 + 3t + 4 = 0$. Thus $t^3 = 2t^2 - 3t - 4$ and so

$$t^4 = t^3 t = (2t^2 - 3t - 4)t = 2t^3 - 3t^2 - 4t$$
$$= 2(2t^2 - 3t - 4) - 3t^2 - 4t = t^2 - 10t - 8.$$

Therefore now

$$t^5 = (t^2 - 10t - 8)t = t^3 - 10t^2 - 8t$$
$$= 2t^2 - 3t - 4 - 10t^2 - 8t = -8t^2 - 11t - 4.$$

Similarly t^6 and higher powers of t can be also expressed in this way.

Example 11.12. Set $t = \sqrt{2}$. We shall work over the field $\mathbb{Q}$ of rational numbers. Set $m(X) = X^2 - 2$. We have $m(t) = 0$ and $m(X)$ is irreducible over $\mathbb{Q}$ by the Eisenstein criterion (8.17) with $p = 2$. Hence $m(X)$ is the minimum polynomial of t and t is algebraic of degree 2 over $\mathbb{Q}$. Therefore by 11.10 $[\mathbb{Q}(t):\mathbb{Q}] = 2$ and the elements of $\mathbb{Q}(t)$ can be expressed uniquely in the form $a + b\sqrt{2}$ with $a, b \in \mathbb{Q}$.

Similarly the elements of $\mathbb{Q}(\sqrt{(-5)})$ are uniquely expressible in the form $a + \sqrt{(-5)}b$ with $a, b \in \mathbb{Q}$. ∎

Example 11.13. Working over $\mathbb{Q}$, set $t = i$. We know that t satisfies no linear equation over $\mathbb{Q}$ because $t \notin \mathbb{Q}$. However t is a root of the quadratic equation $X^2 + 1 = 0$. Therefore t is algebraic of degree 2 over $\mathbb{Q}$. Hence $[\mathbb{Q}(t):\mathbb{Q}] = 2$ and the elements of $\mathbb{Q}(t)$ are uniquely expressible in the form $a + bt$ where $a, b \in \mathbb{Q}$. Thus $\mathbb{Q}(i)$ is the field of Gaussian numbers (7.5). ∎

Example 11.14. Choose t to be one of the three numbers in $\mathbb{C}$ such that $t^3 = 2$. Then t is a root of the equation $X^3 - 2 = 0$, and $X^3 - 2$ is irreducible over $\mathbb{Q}$ by the Eisenstein criterion (8.17) with $p = 2$. Therefore $X^3 - 2$ is the minimum polynomial for t over $\mathbb{Q}$. Hence $[\mathbb{Q}(t):\mathbb{Q}] = 3$ and the elements of $\mathbb{Q}(t)$ can be written uniquely in the form $a + bt + ct^2$ with $a, b, c \in \mathbb{Q}$. ∎

Example 11.15. Working over $\mathbb{C}$ set $t = i + \sqrt{2}$ and $m(X) = X^4 - 2X^2 + 9$. It was shown in 11.3(c) that $m(t) = 0$ and that $m(X)$ is irreducible over $\mathbb{Q}$. Therefore t is algebraic of degree 4 over $\mathbb{Q}$ and the elements of $\mathbb{Q}(t)$ can be written uniquely in the form $a + bt + ct^2 + dt^3$ for some $a, b, c, d \in \mathbb{Q}$. ∎

Example 11.16. Working over the field $\mathbb{R}$ of real numbers, set $t = i$. As in 11.13 we know that t satisfies a quadratic equation but no linear equation over $\mathbb{R}$. Therefore $[\mathbb{R}(i):\mathbb{R}] = 2$ and the elements of $\mathbb{R}(i)$ can be uniquely written in the form $a + ib$ for some $a, b \in \mathbb{R}$. Note that, in fact, $\mathbb{R}(i) = \mathbb{C}$. ∎

Notation 11.17. Let K be an extension of a field F and let $a_1, \ldots, a_r \in K$. Then $F(a_1, \ldots, a_r)$ will denote the smallest subfield of K containing F and the elements $a_1, \ldots, a_r$. That is $F(a_1, \ldots, a_r)$ is the intersection of all the subfields of K containing F and $a_1, \ldots, a_r$.

Comments 11.18

(a) In practice we shall only consider $F(a_1, \ldots, a_r)$ when all the a_i are algebraic over F. In that case we can use 11.9 and induction on r to show that $F(a_1, \ldots, a_r)$ consists of all elements of K which can be written as polynomial expressions in the a_i with coefficients in F, i.e., anything which can be expressed algebraically in terms of the a_i and elements of F.

(b) Since $\mathbb{Q}(i, \sqrt{2})$ is the smallest field which contains $\mathbb{Q}$, i and $\sqrt{2}$, we can also regard $\mathbb{Q}(i, \sqrt{2})$ as being the smallest subfield of $\mathbb{C}$ which contains $\mathbb{Q}(i)$ and $\sqrt{2}$. Thus denoting $F = \mathbb{Q}(i)$ we have $\mathbb{Q}(i, \sqrt{2}) = F(\sqrt{2})$. Similarly if $K = \mathbb{Q}(\sqrt{2})$ then $\mathbb{Q}(i, \sqrt{2}) = K(i)$. Roughly speaking, $\mathbb{Q}(i, \sqrt{2})$ consists of everything which can be expressed in terms of rational numbers, i, and $\sqrt{2}$.

Example 11.19. We shall show that $\mathbb{Q}(i, \sqrt{2}) = \mathbb{Q}(i + \sqrt{2})$ and also that $[\mathbb{Q}(i, \sqrt{2}):\mathbb{Q}] = 4$. We have $i + \sqrt{2} \in \mathbb{Q}(i, \sqrt{2})$ since $i \in \mathbb{Q}(i, \sqrt{2})$ and $\sqrt{2} \in \mathbb{Q}(i, \sqrt{2})$. Set $t = i + \sqrt{2}$. Thus $\mathbb{Q}(i, \sqrt{2})$ is a field containing both $\mathbb{Q}$ and t and so $\mathbb{Q}(i, \sqrt{2}) \supseteq \mathbb{Q}(t)$. In principle it is possible to write both i and $\sqrt{2}$ in terms of rational numbers and powers of t, and that would show that $\mathbb{Q}(t) \supseteq \mathbb{Q}(i, \sqrt{2})$. For instance $2 = (t - i)^2 = t^2 - 2it - 1$ so that $i = (t^2 - 3)/2t$.

Alternatively we can proceed as follows: We have $\mathbb{Q} \subseteq \mathbb{Q}(t) \subseteq \mathbb{Q}(i, \sqrt{2})$. Hence by 9.24 we have $[\mathbb{Q}(i, \sqrt{2}):\mathbb{Q}] = [\mathbb{Q}(i, \sqrt{2}):\mathbb{Q}(t)][\mathbb{Q}(t):\mathbb{Q}]$. Now we have found the minimum polynomial for t over $\mathbb{Q}$ in 11.15 and hence shown that $[\mathbb{Q}(t):\mathbb{Q}] = 4$. In a moment we shall show that $[\mathbb{Q}(i, \sqrt{2}):\mathbb{Q}] = 4$. Hence $4 = [\mathbb{Q}(i, \sqrt{2}):\mathbb{Q}(t)] \cdot 4$ and so $[\mathbb{Q}(i, \sqrt{2}):\mathbb{Q}(t)] = 1$. Therefore by Question 11, Exercises 9 we have $\mathbb{Q}(i, \sqrt{2}) = \mathbb{Q}(t)$.

It remains to show that $[\mathbb{Q}(i, \sqrt{2}) : \mathbb{Q}] = 4$. By 11.12 we know that $[\mathbb{Q}(\sqrt{2}) : \mathbb{Q}] = 2$. Set $F = \mathbb{Q}(\sqrt{2})$. The coefficients of the polynomial $X^2 + 1$ certainly belong to F, and i is a root of the equation $X^2 + 1 = 0$. Thus i satisfies a quadratic equation with coefficients in F. The elements of F are real numbers because they can be expressed in terms of rational numbers and $\sqrt{2}$. Hence $i \notin F$. Therefore i satisfies no linear equation with coefficients in F. Hence $X^2 + 1$ is the minimum polynomial for i over F and so $2 = [F(i) : F] = [\mathbb{Q}(i, \sqrt{2}) : \mathbb{Q}(\sqrt{2})]$. Therefore $[\mathbb{Q}(i, \sqrt{2}) : \mathbb{Q}] = [\mathbb{Q}(i, \sqrt{2}) : \mathbb{Q}(\sqrt{2})][\mathbb{Q}(\sqrt{2}) : \mathbb{Q}] = 2 \cdot 2 = 4$ as required.　∎

Lemma 11.20.　*Let K be an extension of a field F and let $a, b \in K$ such that a and b are algebraic over F. Then $F(a, b)$ is finite dimensional as a vector space over F.*

Proof.　Let a be algebraic of degree n over F. Then by 11.10 we have $[F(a) : F] = n$. Let $m(X)$ be the minimum polynomial for b over F with $\partial(m) = r$. The coefficients of $m(X)$ are in F, and $F \subseteq F(a)$. Therefore the coefficients of $m(X)$ are in $F(a)$, and also $m(b) = 0$. Thus b is a root of a polynomial equation of degree r with coefficients in $F(a)$. Hence b is algebraic of degree k over $F(a)$ for some positive integer $k \leq r$ (and, in general, k can be strictly smaller than r). Therefore $[F(a, b), F(a)] = k$ and so $[F(a, b) : F] = [F(a, b) : F(a)][F(a) : F] = kn$.　∎

Lemma 11.21.　*Let K be a finite extension of a field F and let $t \in K$. Then t is algebraic over F.*

Proof.　We have $[K : F] = n$ for some positive integer n. Thus K has dimension n as a vector space over F. Hence the $n + 1$ elements $1, t, t^2, \ldots, t^n \in K$ are linearly dependent over F. Therefore there are elements $a_0, a_1, \ldots, a_n \in F$, not all zero, such that

$$a_0 + a_1 t + \cdots + a_n t^n = 0.$$

Thus t satisfies a non-trivial equation over F.　∎

Theorem 11.22.　*Let K be an extension of a field F and let H be the set of elements of K which are algebraic over F. Then H is a subfield of K.*

Proof.　Clearly H contains the elements 0 and 1 of K. Let $a, b \in H$. By 11.20 we know that $[F(a, b) : F]$ is finite. However, $F(a, b)$ contains a and b and hence contains $a - b$ and ab. Therefore by 11.21, $a - b$ and $ab \in H$. Suppose that $a \neq 0$ and that $f_0 + f_1 a + \cdots + f_r a^r = 0$ for some $f_i \in F$. Then $f_r + f_{r-1}(1/a) + \cdots + f_0(1/a)^r = 0$, so that $1/a$ is algebraic over F. Therefore $1/a \in H$. Thus H is a subfield of K.　∎

Definition 11.23. The field H defined above is called the *algebraic closure* of F in K.

Corollary 11.24. *The set of algebraic numbers is a subfield of* $\mathbb{C}$.

Proof. Take $K = \mathbb{C}$ and $F = \mathbb{Q}$ in 11.22. ∎

Exercises 11

1. Show that a number t is algebraic of degree 1 over the field F if and only if $t \in F$.

2. Give an example of a field F and elements a and b which are algebraic over F such that $F(a, b) \neq F(a + b)$.

3. Let K be an extension of a field F and let $t \in K$. Prove that $F(t) = F(a + bt)$ for all $a, b \in F$ with $b \neq 0$.

4. Let K be a *quadratic extension* of the field F (i.e. $[K : F] = 2$). Suppose that the characteristic of F is not 2. Prove that $K = F(\sqrt{d})$ for some $d \in F$. (*Hint:* Fix an element $t \in K$ such that $t \notin F$. Use the fact that the three elements $1, t, t^2$ are linearly dependent over F to show that t satisfies a quadratic equation over F. Solve the equation. The fact that F has characteristic other than 2 makes it possible to use the standard formula for solving quadratic equations.)

5. You are given that there is a field K of order 4 and that K has a subfield F of order 2. Prove that there is no element $d \in F$ such that $K = F(\sqrt{d})$.

6. Set $t = \sqrt{2} + \sqrt{3}$. Find the minimum polynomial for t over $\mathbb{Q}$ and express t^5 in the form $a + bt + ct^2 + dt^3$ for some $a, b, c, d \in \mathbb{Q}$.

7. Let $f(X)$ be a monic polynomial of degree n and suppose that

$$f(X) = (X - a_1)(X - a_2) \cdots (X - a_n).$$

Express $a_1 + a_2 + \cdots + a_n$ and $a_1 a_2 \cdots a_n$ in terms of the coefficients of $f(X)$. Note that under suitable conditions this gives expressions for the sum and product of the roots of a polynomial equation in terms of the coefficients.

8. Let K be an extension of a field F and let $f(X)$ be a monic irreducible cubic polynomial with coefficients in F. Suppose

that the equation $f(X) = 0$ has three distinct roots $a, b, c \in K$. Thus $f(X) = (X - a)(X - b)(X - c)$. Prove each of the following statements:

  (i) $[F(a) : F] = 3$;

  (ii) b satisfies a quadratic equation with coefficients in $F(a)$;

  (iii) $F(a, b, c) = F(a, b)$ (*Hint:* Use question 7);

  (iv) $[F(a, b, c) : F] = 3$ or 6.

9. Working over $\mathbb{Q}$ set $f(X) = X^3 + X^2 - 2X - 1$. Prove that $f(X)$ is irreducible over $\mathbb{Q}$. Let r be a root of the equation $f(X) = 0$. Prove that $r^2 - 2$ is also a root of the same equation and that $r^2 - 2 \neq r$. Hence show that $F(r)$ contains all three roots of the equation $f(X) = 0$.

10. (Formula for solving cubic equations) Let $f(X)$ be a cubic polynomial with coefficients in a field F whose characteristic is neither 2 nor 3. Without loss of generality we may suppose that $f(X)$ is monic. Suppose that $f(X) = X^3 + aX^2 + bX + c$. Set $Y = X + a/3$. Show that $f(X)$ is of the form $Y^3 + pY + q$. Set $Y = W - p/3W$. Verify that $Y^3 + pY + q = W^3 + q - p^3/27W^3$. Show that the roots of the equation $W^3 + q - p^3/27W^3 = 0$ are of the form

$$W = \left\{ -\tfrac{1}{2}q + \tfrac{1}{2}(q^2 + 4p^3/27)^{1/2} \right\}^{1/3}$$

for various choices of the square and cube roots. Since $Y = W - p/3W$, this gives a formula for the roots of the equation $Y^3 + pY + q = 0$ in terms of p and q.

11. Let F be a finite field with p^n elements for some prime p and positive integer n. Let t be a generator for the cyclic group of non-zero elements of F under multiplication and let S be the subfield of F with $|S| = p$. Prove that $F = S(t)$ and that t is algebraic of degree n over S.

12 Ruler-and-compass constructions

In the Greek tradition of geometry there has been much interest in determining which geometrical constructions can be performed using only two very simple devices. This question is asked in an idealized context where constructions can be performed with complete accuracy. The first permitted device is a ruler or straight edge; this makes it possible to draw the line joining distinct points A and B in the plane and to extend the line as far as necessary in either direction. The ruler has no markings on it and it cannot be used for measuring lengths. The second device is a pair of compasses; this makes it possible to draw the circle with centre at given point A passing through a second point B. It was known to the ancient Greek geometers that it is possible, using only these devices in combination, to perform such constructions as bisecting a given angle or dropping a perpendicular from a point to a line (details below). The Greek geometers wanted to be able to trisect an arbitrary angle and to do certain other constructions. Some of these including the trisection of an angle they failed to achieve. It was only in the 19th century AD that it was conclusively shown that some of these constructions are impossible under the constraints given. We shall discuss this theory in this chapter, give some examples of constructions which can be performed, and prove that certain others are impossible. In particular, we determine the positive integers n for which it is possible to construct the regular n-sided polygon.

Example 12.1 (Bisecting an angle).   We shall explain how to bisect the angle between the lines OA and OB as in Fig. 12.1. Draw the circle centred at O passing through A. Let it cut OB at A'. Thus OA and OA' have the same length. Let O and X be the two points of intersection of the circle centred at A passing through O with the circle centred at A' passing through O. These circles have radii of the same length, so that AX and $A'X$ are of equal length. The triangles OAX and $OA'X$ are identical in size, so that the angle between OA and OX equals the angle between OA' and OX. Therefore OX bisects the given angle. ■

Example 12.2 (Bisecting a line-segment).   We shall show how to find the mid-point of the line joining the distinct points A and B (Fig. 12.2). Let X and Y be the points of intersection of the circle centred at A through B with the circle centred at B through A. The desired point is where the lines XY and AB meet. ■

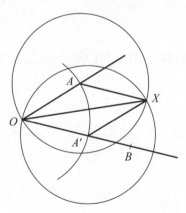

Fig. 12.1. Bisecting an angle

Example 12.3 (Dropping a perpendicular).   Let P be a point which is not on the line AB. We shall show how to construct the line through P at right angles to the line AB (Fig. 12.3). Draw the circle centred at P passing through A and let it cut AB again at A'. Construct the mid-point Q of AA' as in 12.2. Then PQ is the desired line.   ∎

Example 12.4 (Constructing a parallel).   Let P be a point which is not on the line AB. We shall show how to construct the line through P which is parallel to AB (Fig. 12.4). As is 12.3 construct the point Q on AB such that PQ is at right angles to AB. Choose one of the points R where AB cuts the circle centred at Q passing through P. Thus QR and QP have the same length. Let S be the point other than Q where the circle centred at P passing through Q cuts the circle centred at R passing through Q. Then $PQRS$ is a square and PS is parallel to AB.   ∎

Comments 12.5

(a) Many other constructions can be added to those given in 12.1–12.4.

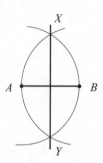

Fig. 12.2. Bisecting a line-segment

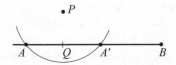

Fig. 12.3. Dropping a perpendicular

For example, given distinct points A and B it is possible to construct the line through B at right angles to AB (see Question 4, Exercises 12).

(b) From now on it will often be convenient to think of the plane as having a fixed Cartesian (x, y)-coordinate system as usual. The point $(1, 0)$ is thought of as a random point marked on the x-axis and is used to define a unit of length. Thus the line segment joining $(0, 0)$ and $(1, 0)$ is defined to have length 1. All other lengths in the plane are based on this length in the usual way. Suppose that we are given distinct points A and B in the plane such that neither point is on the x-axis. It is possible, as follows, to find a point P on the x-axis such that OP and AB have the same length. In this manner we can transfer a marked length from one line to another. Let C be a point where the line through A parallel to the x-axis cuts the circle centred at A passing through B. Thus AC and AB have the same length. Let P be the point where the line through C parallel to AO cuts the x-axis.

Definition 12.6. A positive real number a is said to be *constructible* if, starting from the points $(0, 0)$ and $(1, 0)$, it is possible to construct a line-segment of length a using ruler and compasses only.

If a is a negative real number then a is considered to be constructible if the positive number $-a$ is constructible.

The real number 0 is defined to be constructible.

Comments 12.7. Initially the only real numbers which we know to be

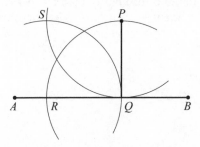

Fig. 12.4. Constructing a parallel

constructible are 0, 1, and -1. We shall show that all rational numbers are constructible, and that the set of all constructible numbers is a field which contains the square roots of all its positive elements. We shall concentrate on the positive constructible numbers and leave the reader to check the negative cases. The word 'constructible' in this context will always mean 'constructible in a finite number of steps using only ruler and compasses'.

Lemma 12.8. *The positive integers are constructible.*

Proof. Let OA be a line-segment of length 1. Let B be the point other than O where the line OA cuts the circle centred at A passing through 0. Thus A is the mid-point of OB and OB has length 2. Hence 2 is constructible. Let C be the point other than A where OB cuts the circle centred at B passing through A. Then OC has length 3. By continuing in this way we can construct any positive integer.   ■

Lemma 12.9. *Let a and b be constructible positive real numbers. Then the numbers $a + b$, ab, a/b, $\sqrt{a}$ are constructible.*

Proof. We leave $a + b$ to the reader. To construct ab we start with a line segment OA of length a and a point P not on the line OA (Fig. 12.5). We construct the points U and B on OP such that U and B are on the same side of O and such that OU and OB have lengths 1 and b respectively. Let the line through B parallel to UA cut OA at D and let x be the length of OD. The triangles OAU and ODB have equal corresponding angles, so that they have the same shape and differ only by a scaling factor in size. Therefore the ratios of the lengths of corresponding sides are equal. In particular $OA/OD = OU/OB$. Thus $a/x = 1/b$ and so $x = ab$. Hence ab is constructible.

In order to construct a/b we start as in Fig. 12.5 but this time take D to be the point where the line through U parallel to BA cuts OA (Fig. 12.6). The same type of argument gives $x/a = 1/b$. Therefore $x = a/b$.

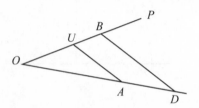

Fig. 12.5. Proving that ab is constructible

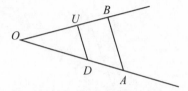

Fig. 12.6. Proving that a/b is constructible

Finally, to construct $\sqrt{a}$ start with a line segment OA of length a (Fig. 12.7). Let P be the point on OA on the opposite side of O from A and such that OP has length 1. Let X be the mid-point of PA. Draw the circle centred at X and passing through P and A. This circle has PA as a diameter. Construct the line through O perpendicular to OA (see Question 4, Exercises 12). Let Q be one of the points where this line meets the circle. Now let x be the length of OQ. It is a standard theorem of Euclidean geometry that a diameter of a circle subtends an angle of 90° at a point on the circumference. Hence QP and QA meet at right angles. From this it follows that the right-angled triangles POQ and QOA have equal corresponding angles. Therefore $PO/OQ = OQ/OA$. Thus $1/x = x/a$ and so $x = \sqrt{a}$.   ■

Corollary 12.10.   *The set of constructible real numbers is a field which contains all rational numbers and the square roots of its positive elements.*

Proof.   This is a straightforward consequence of 12.9.   ■

Proposition 12.11.   *The regular pentagon is constructible.*

Proof.   Working in the standard Cartesian coordinate system take $O = (0,0)$ and $A = (1,0)$ (Fig. 12.8). Draw the circle centred at O passing

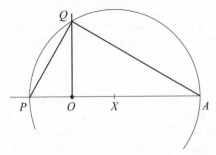

Fig. 12.7. Proving that $\sqrt{a}$ is constructible

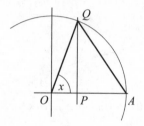

Fig. 12.8. Proving that the regular pentagon is constructible

through A. We call this the unit circle. By 12.10 we can construct the point P between O and A such that OP has length $(\sqrt{5} - 1)/4$. Let the line through P at right angles to OA cut the unit circle at Q and let x be the angle between OP and OQ. The right-angled triangle OPQ has base OP of length $(\sqrt{5} - 1)/4$ and hypotenuse OQ of length 1. Therefore $\cos x = (\sqrt{5} - 1)/4$ so that $x = 72°$ (see Question 10, Exercises 12). Thus AQ subtends an angle of $72°$ at O, and AQ is one edge of a regular pentagon inscribed in the unit circle. We leave it as an exercise to show how to construct the complete pentagon. ∎

Comments 12.12

(i) We shall return later in this chapter to a more detailed study of constructibility of regular polygons.

(ii) We showed in 12.9 how to construct square roots. The next aim, roughly speaking, is to show that the only new numbers which can be constructed from given ones are square roots.

Theorem 12.13. *Let $P = (x, y)$ be a point in the plane. Suppose that (x, y) can be constructed, starting from the points $(0,0)$ and $(1,0)$, in a finite number of steps using only ruler and compasses. Then there is a field F such that $x, y \in F$ and $[F : \mathbb{Q}] = 2^n$ for some non-negative integer n. In fact the coordinates x and y of P can be obtained from elements of $\mathbb{Q}$ by taking a finite number of square roots. (For example, x and y might be $5 - \sqrt{(1 + \sqrt{7})}$ and $10\sqrt{(\sqrt{(6 + \sqrt{3})} - \sqrt{5})}$.)*

Proof. The idea of the proof is as follows: suppose that at an intermediate stage in the construction we have constructed a set S of points. Let K be the smallest field which contains all the coordinates of points in S. Now let X be a point which is constructed from points in S by a single basic ruler-and-compasses construction, and let L be the smallest field which contains K and the coordinates of X. We shall show that either the coordinates of X are in K, in which case $L = K$ and $[L : K] = 1$, or the coordinates of X involve a 'new' square root of an element of K and $[L : K] = 2$. Thus in order to construct P we start from points $(0, 0)$ and $(1, 0)$ with coordinates in $\mathbb{Q}$ and construct a finite increasing sequence of fields $\mathbb{Q} = F_0 \subseteq F_1 \subseteq F_2 \subseteq \cdots \subseteq F_r$ such that F_r contains the coordinates of P and for each i we have $[F_i : F_{i-1}] = 1$ or 2. Set $F = F_r$. Since, by 9.24,

$$[F : \mathbb{Q}] = [F_r : \mathbb{Q}] = [F_r : F_{r-1}][F_{r-1} : F_{r-2}] \cdots [F_1 : \mathbb{Q}],$$

we have $[F : \mathbb{Q}] = 2^n$ for some non-negative integer n.

Thus we may now concentrate on the following situation: S is a given set of points in the plane; K is the smallest field which contains the coordinates of all points in S; $P = (s, t)$ is a point which is constructible from points in S in one step as explained below; L is the smallest field which contains K and the coordinates of P, i.e. $L = K(s, t)$. We must show that $[L : K] = 1$ or 2. There are only three ways of constructing P in one step.

(i) *Line meets line.* We start with distinct points $A, B, C, D \in S$ and construct P as the point of intersection of the lines AB and CD. The coordinates of A and B are in the field K. Therefore the equation of the line AB can be written in the form $ux + vy + w = 0$ with $u, v, w \in K$. Similarly the equation of CD can be written in the form $u'x + v'y + w' = 0$ for some $u', v', w' \in K$. Therefore the coordinates s and t of P are solutions of the simultaneous equations

$$ux + vy + w = 0 \qquad \text{and} \qquad u'x + v'y + w' = 0.$$

Hence $s, t \in K$ so that $L = K(s, t) = K$.

(ii) *Line meets circle.* We take points $A, B, C, D \in S$ and construct P as one of the points where the line AB cuts the circle centred at C passing through D. The equation of the line AB can be written in the form $ux + vy + w = 0$ with $u, v, w \in K$. We have $C = (c, c')$ and $D = (d, d')$ for some $c, c', d, d' \in K$. Hence the equation of the circle centred at C passing through D is

$$(x - c)^2 + (y - c')^2 = (d - c)^2 + (d' - c')^2,$$

which can be written in the form $x^2 + y^2 + fx + gy + h = 0$ with $f, g, h \in K$. We cannot have both u and v zero. Suppose that $v \neq 0$. Then we can use $ux + vy + w = 0$ to write y in terms of x. By substituting for y in the equation of the circle we obtain a quadratic equation for x with coefficients in K. Since s is one solution of this quadratic equation we have (using the formula for solving quadratic equations) $s = a + b \cdot \sqrt{k}$ for some

$a, b, k \in K$. Thus $s \in K(\sqrt{k})$. Since $us + vt + w = 0$ we also have $t \in K(\sqrt{k})$. (The case $v = 0$ needs to be treated separately and we leave it to the reader). Therefore $L = K(s, t) = K(\sqrt{k})$. If $\sqrt{k} \notin K$ then $[L : K] = [K(\sqrt{k}) : K] = 2$. If $\sqrt{k} \in K$, i.e. if k is a perfect square in K, then $L = K$ and $[L : K] = 1$.

(iii) *Circle meets circle*. We construct the point P as a point of intersection of two circles given by the equations

$$x^2 + y^2 + fx + gy + h = 0 \qquad \text{and} \qquad x^2 + y^2 + f'x + g'y + h' = 0$$

with $f, g, h, f', g', h' \in K$. To find the points of intersection of these circles we need only solve the equations

$$x^2 + y^2 + fx + gy + h = 0 \qquad \text{and} \qquad (f - f')x + (g - g')y + h - h' = 0.$$

This case was dealt with in (ii).   ∎

Comments 12.14. We can use 12.13 and the theory of field extensions and algebraic elements as follows to show that certain constructions are not possible. We suppose that the construction is possible and aim to derive a contradiction. We construct a point one of whose coordinates t is algebraic over $\mathbb{Q}$ of degree n, say. We show that n is not a power of 2; thus for example, n might be 3. We know from 12.13 that there is a field F containing t such that $[F : \mathbb{Q}] = 2^k$ for some non-negative integer k. Thus we have $\mathbb{Q} \subseteq \mathbb{Q}(t) \subseteq F$ and so by 9.24, $[\mathbb{Q}(t) : \mathbb{Q}] \mid [F : \mathbb{Q}]$, i.e. $n \mid 2^k$. This is the desired contradiction because we have earlier shown that n is not a power of 2. This method, which dates from the 19th century, was used to solve three problems of classical geometry. We present them as the next three theorems.

Theorem 12.15 (The impossibility of trisection of an angle). *It is impossible to trisect an arbitrary angle by the ruler-and-compass method.* (Here 'trisect' means 'divide into three equal parts'.)

Proof. We know that it is possible to construct an angle of 60° (Question 1, Exercises 12). Suppose that we have a ruler-and-compasses construction which can trisect this angle. Then we can construct an angle of 20°. We now work with the trigonometric functions sine and cosine. We start from the points $O = (0, 0)$ and $A = (1, 0)$ and construct the usual unit circle. We are assuming that we can construct the line through O going up to the right making an angle of 20° with OA. Let this line cut the unit circle at P. Then $P = (\cos 20, \sin 20)$. For later convenience we set $t = 2 \cos 20$. As was shown in 12.14 we know that $t \in F$ where F is a field such that $[F : \mathbb{Q}]$ is a power of 2 and hence $[\mathbb{Q}(t) : \mathbb{Q}]$ is a power of 2. We shall obtain the desired contradiction by showing that $[\mathbb{Q}(t) : \mathbb{Q}] = 3$.

We have by de Moivre's theorem

$$(\cos 20 + i \sin 20)^3 = \cos 60 + i \sin 60 = (1 + i\sqrt{3})/2. \tag{i}$$

However, expanding $(\cos 20 + i \sin 20)^3$ by the binomial theorem we have

$$(\cos 20 + i \sin 20)^3 = \cos^3 20 + 3i \cos^2 20 \sin 20 - 3 \cos 20 \sin^2 20 - i \sin^3 20. \tag{ii}$$

Equating the real parts of (i) and (ii) we have

$$\tfrac{1}{2} = \cos^3 20 - 3 \cos 20 (1 - \cos^2 20)$$
$$= 4 \cos^3 20 - 3 \cos 20 = (t^3 - 3t)/2.$$

Thus $t^3 - 3t = 1$ and so t is a root of the equation $f(X) = 0$ where $f(X) = X^3 - 3X - 1$. Suppose that $f(X)$ is not irreducible over $\mathbb{Q}$. Then by 8.21 there is an integer a such that $f(a) = 0$ and a divides the constant term of $f(X)$. This is a contradiction because $f(1) \neq 0$ and $f(-1) \neq 0$. Therefore $f(X)$ is irreducible over $\mathbb{Q}$ and so is the minimum polynomial for t over $\mathbb{Q}$. Hence $[\mathbb{Q}(t):\mathbb{Q}] = \partial(f) = 3$. ∎

Theorem 12.16 (The impossibility of doubling a cube). *Given a cube it is impossible to construct one of double the volume by ruler-and-compasses construction.*

Proof. Take the length of a side of the given cube as the unit length. In order to double the volume we need to construct a segment of length $2^{1/3}$. As in 12.14 if such a length were constructible then we would have that $[\mathbb{Q}(2^{1/3}):\mathbb{Q}]$ is a power of 2. However, we know by 11.14 that $[\mathbb{Q}(2^{1/3}):\mathbb{Q}] = 3$. Thus it is impossible to construct a cube of double volume. ∎

Theorem 12.17 (The impossibility of squaring the circle). *Given an arbitrary circle, it is impossible to construct a square with the same area as the circle.*

Proof. We can certainly construct a circle with unit radius. This circle has area π. Thus we need to construct a square with sides $\sqrt{\pi}$. Now it was shown by Lindemann in 1882 that π is not algebraic over $\mathbb{Q}$ (see Stewart (1989), p. 66). So by 11.21 $[\mathbb{Q}(\sqrt{\pi}):\mathbb{Q}]$ cannot be finite. So as in 12.14, the length $\sqrt{\pi}$ is not constructible. ∎

Conventions 12.18. Let n be a positive integer such that $n \neq 1$ or 2. By a 'regular n-gon' we mean the regular n-sided polygon, i.e. the polygon in the plane with n sides of equal length meeting at equal angles.

p will always denote an odd prime in the rest of this chapter.

Definition 12.19. For each non-negative integer n the numbers $F_n = 2^{2^n} + 1$ are known as the *Fermat numbers*. A prime number p is called a *Fermat prime* if $p = F_n$ for some n.

Comments 12.20. We have $F_0 = 3$, $F_1 = 5$. $F_2 = 17$, $F_3 = 257$, $F_4 = 65\,537$, all of which are prime. Fermat had conjectured that F_n is always prime. However, Euler showed that $F_5 = 2^{32} + 1$ is divisible by 641. It is now known that none of the numbers $F_5, F_6, \ldots, F_{16}$ is a prime. The reason why we have introduced Fermat primes here is partly revealed by the next result, which shows that if the regular p-gon is constructible, then p is a Fermat prime. Thus if p is a prime but not a Fermat prime, e.g. if $p = 7$, 11, or 13, then the regular p-gon is not constructible. The main aim of the rest of this chapter is to show that a regular n-gon is constructible if and only if n is of the form 2^k or $2^k p_1 p_2 \cdots p_r$, where k is a non-negative integer, r is a positive integer and the p_i are distinct Fermat primes. It was not until the 1790s that Gauss made the first significant advance in this area since Greek times, by showing how to construct the regular 17-gon.

Proposition 12.21. *Suppose that the regular p-gon is constructible. Then p is a Fermat prime.*

Proof. All the angles will be measured in degrees. Each side of the regular p-gon subtends an angle of $360/p$ degrees at the centre. Hence, by assumption, we can construct an angle of $360/p$ degrees. Set $O = (0,0)$ and $A = (1,0)$. Let P be the point where the unit circle cuts the line going up to the right from O at an angle of $360/p$ degrees to OA. The x-coordinate of P is t, where $t = \cos(360/p)$. Therefore t is a constructible number and by 12.14, $[\mathbb{Q}(t):\mathbb{Q}] = 2^k$ for some integer k.

Set $w = e^{2i\pi/p}$. Then $t = \cos(360/p) = (w + 1/w)/2$. Hence $w + 1/w = 2t$ so that $w^2 - 2tw + 1 = 0$. Thus w is a root of the quadratic equation $X^2 - 2tX + 1 = 0$ which has coefficients in $\mathbb{Q}(t)$. Also $w \notin \mathbb{Q}(t)$ since $\mathbb{Q}(t) \subseteq \mathbb{R}$ but $w \notin \mathbb{R}$. Therefore w is algebraic of degree 2 over $\mathbb{Q}(t)$ and $[\mathbb{Q}(t,w):\mathbb{Q}(t)] = 2$. However, $t \in \mathbb{Q}(w)$ since $t = (w + 1/w)/2$. Hence $\mathbb{Q}(t,w) = \mathbb{Q}(w)$ and so $[\mathbb{Q}(w):\mathbb{Q}(t)] = 2$. Therefore by 9.24, $[\mathbb{Q}(w):\mathbb{Q}] = [\mathbb{Q}(w):\mathbb{Q}(t)][\mathbb{Q}(t):\mathbb{Q}] = 2^{k+1}$.

We shall next show that $[\mathbb{Q}(w):\mathbb{Q}] = p - 1$. It will then follow that $p = 2^{k+1} + 1$ and that p is a Fermat prime (see Question 5, Exercises 12). Set

$$m(X) = X^{p-1} + X^{p-2} + \cdots + X^2 + X + 1.$$

We have

$$m(w) = w^{p-1} + w^{p-2} + \cdots + w + 1 = (w^p - 1)/(w - 1) = 0$$

because $w^p = 1$. Now by 8.29 $m(X)$ is irreducible over $\mathbb{Q}$. Hence $m(X)$ is the minimum polynomial for w over $\mathbb{Q}$. It follows by 11.10 that $[\mathbb{Q}(w):\mathbb{Q}] = \partial(m) = p - 1$. ∎

Proposition 12.22. *Suppose that the regular n-gon is constructible. Let p be an odd prime factor of n. Then p is a Fermat prime and p^2 does not divide n.*

Proof. As the regular n-gon is constructible, we can construct a regular p-gon by joining some of the vertices of the n-gon. For example if we have a regular 28-gon say, with vertices $P_1, P_2, \ldots, P_{28}$ in cyclic order, then we can form a regular 7-gon by using the vertices $P_1, P_5, P_9, P_{13}, P_{17}, P_{21}, P_{25}$. Therefore by 12.21 p is a Fermat prime.

Suppose now that $p^2 \mid n$. As above we can construct a regular p^2-gon by joining some of the vertices of the regular n-gon. Hence we can construct an angle of $360/p^2$ and, as shown in the proof of 12.21, also the real number $t = \cos(360/p^2)$. Set $w = e^{2i\pi/p^2}$. Then $t = (w + 1/w)/2$. As in the proof of 12.21 we have $[\mathbb{Q}(w):\mathbb{Q}] = 2^{k+1}$ for some integer $k \geq 0$. We shall show that $[\mathbb{Q}(w):\mathbb{Q}] = p(p-1)$. It will then follow that $p(p-1) = 2^{k+1}$, which is a contradiction since p is odd.

It remains to show that $[\mathbb{Q}(w):\mathbb{Q}] = p(p-1)$. Note that $w^{p^2} = 1$ but $w^p \neq 1$. Set

$$m(X) = (X^{p^2} - 1)/(X^p - 1) = ((X^p)^p - 1)/(X^p - 1)$$

$$= (X^p)^{p-1} + (X^p)^{p-2} + \cdots + X^p + 1.$$

Then $m(w) = (w^{p^2} - 1)/(w^p - 1) = 0$. By 8.30 we know that $m(X)$ is irreducible over $\mathbb{Q}$. So it follows by 11.10 that $[\mathbb{Q}(w):\mathbb{Q}] = \partial(m) = p(p-1)$. ∎

Proposition 12.23. *Suppose that the regular a-gon and the regular b-gon are constructible where a and b are relatively prime positive integers. Then the regular ab-gon is constructible.*

Proof. By 3.5 there are integers u and v such that $au + bv = 1$. Hence

$$360/ab = u \cdot 360/b + v \cdot 360/a.$$

Note that one of u and v is positive and the other negative. We know that we can construct angles of $360/a$ and $360/b$ degrees. By adding these angles to themselves the right number of times in a positive or negative sense, we can construct an angle of $360/ab$ degrees. ∎

Theorem 12.24. *Recall our convention that $n \geq 3$. The regular n-gon is constructible if and only if $n = 2^k$ or $2^k p_1 p_2 \cdots p_r$ where k is a non-negative integer, r is a positive integer and the p_i are distinct Fermat primes.*

Proof. If the regular n-gon is constructible then n has the right form by 12.22.

Suppose conversely that $n = 2^k$ or $2^k p_1 p_2 \cdots p_r$ as above. We wish to show that the regular n-gon is constructible. By 12.23 it is enough to show that the regular 2^k-gon and each regular p_i-gon are constructible. We leave the regular 2^k-gon as an exercise (Question 6, Exercises 12).In 12.25 we shall show how to construct the regular 17-gon, and the same method can be applied to show that the regular p-gon is constructible for every Fermat prime p. ∎

Theorem 12.25 (Gauss). *The regular 17-gon is constructible.*

Proof. Set $w = e^{i\pi/17}$. Then $w^{17} = 1$ and $w \neq 1$. Set

$$m(X) = (X^{17} - 1)/(X - 1) = X^{16} + X^{15} + \cdots + X + 1.$$

Then $m(w) = 0$ and by 8.29 $m(X)$ is irreducible over $\mathbb{Q}$. Hence $m(X)$ is the minimum polynomial for w over $\mathbb{Q}$ and so by 11.10 $[\mathbb{Q}(w):\mathbb{Q}] = \partial(m) = 16$.

We first outline the proof. We shall find a subfield K of $\mathbb{Q}(w)$ such that $[\mathbb{Q}(w):K] = 2$ and $\mathbb{Q}(w) = K(\sqrt{k})$ for some $k \in K$; then a subfield L of K such that $[K:L] = 2$ and $K = L(\sqrt{l})$ for some $l \in L$; then a subfield M of L such that $[L:M] = 2$ and $L = M(\sqrt{m})$ for some $m \in M$. The subfield M will also have the property that $[M:\mathbb{Q}] = 2$ and $M = \mathbb{Q}(\sqrt{q})$ for some $q \in \mathbb{Q}$. Thus with this construction we can go up from $\mathbb{Q}$ to $\mathbb{Q}(w)$ in four stages where at each stage we bring in a new square root, firstly $\sqrt{q}$ then $\sqrt{m}$ then $\sqrt{l}$ and finally $\sqrt{k}$. Thus every element of $\mathbb{Q}(w)$ can be expressed in terms of nested square roots starting from rational numbers. We have $\cos(360/17) = (w + 1/w)/2$ where we measure angles in degrees. Hence $\cos(360/17)$ can be expressed in terms of square roots. We have shown in 12.9 that square roots can be constructed. It follows that we can construct a line-segment of length $\cos(360/17)$. Take $O = (0,0)$ and $A = (1,0)$. Mark the point P between O and A such that $P = (\cos(360/17), 0)$. Draw the line through P perpendicular to OA and let Q be the point in the first quadrant where it cuts the unit circle. Then OQ and OA meet at $360/17$ degrees. It is now easy to construct the regular 17-gon with centre O and having AQ as one edge.

Next we explain why the number 3 will be of importance. We have $[\mathbb{Q}(w):\mathbb{Q}] = 16$. Let G be the multiplicative group of non-zero elements of the field $\mathbb{Z}/17\mathbb{Z}$. Then $|G| = 16$. We also know by 10.8 that G is cyclic. In fact it can be shown by direct calculation that 3 is a generator of G. This is why we set $\alpha(w) = w^3$ below.

Since $[\mathbb{Q}(w):\mathbb{Q}] = 16$, the elements $1, w, w^2, \ldots, w^{15}$ form a basis for $\mathbb{Q}(w)$ as a vector space over $\mathbb{Q}$. The elements $w, w^2, \ldots, w^{16}$ also form a basis for this vector space (see Question 7, Exercises 12). Note that $1 + w + w^2 + \cdots + w^{16} = 0$.

We define a function $\alpha: \mathbb{Q}(w) \to \mathbb{Q}(w)$ as follows. We want α to preserve sums and products (i.e. α to be a homomorphism, see 13.1) and to be such that $\alpha(w) = w^3$. The rest of this paragraph is devoted to setting up the function α. Let $z \in \mathbb{Q}(w)$. Then z can be written as a polynomial expression in w with coefficients in $\mathbb{Q}$, i.e. $z = f(w)$ for some polynomial $f(X)$ with coefficients in $\mathbb{Q}$. This representation for z is not unique because we have deliberately not placed any restriction on the degree of $f(X)$. Set $\alpha(z) = f(w^3)$. Thus, for example, if

$$z = 10w^3 - w^{15}/2 + 19w^{20} - w^{50}/4$$

then

$$\alpha(z) = 10w^9 - w^{45}/2 + 19w^{60} - w^{150}/4.$$

In other words, write z suitably in terms of w and then replace w by w^3. We must first show that α is well defined. We suppose that we have two polynomials $f(X)$ and $g(X)$ with coefficients in $\mathbb{Q}$ such that $z = f(w) = g(w)$. We must show that $f(w^3) = g(w^3)$, so that the value of $\alpha(z)$ is the same whether we use $z = f(w)$ or $z = g(w)$. Since $f(w) - g(w) = 0$ we know that w is a root of the equation $f(X) - g(X) = 0$ where $f(X) - g(X)$ is a polynomial with coefficients in $\mathbb{Q}$. However, as shown earlier in the proof, $m(X)$ is the minimum polynomial for w over $\mathbb{Q}$. Therefore by 11.6 $m(X) \mid f(X) - g(X)$. Hence $f(X) - g(X) = m(X)h(X)$ for some polynomial $h(X)$ with coefficients in $\mathbb{Q}$. Since $m(X) = (X^{17} - 1)/(X - 1)$ we have $m(w^3) = (w^{51} - 1)/(w^3 - 1) = 0$, noting that $w^3 \neq 1$. Hence

$$f(w^3) - g(w^3) = m(w^3)h(w^3) = 0$$

as required. Therefore α is well defined. As α acts by substituting w^3 for w, it is easy to check that for all $z, z' \in \mathbb{Q}(w)$ and $q \in \mathbb{Q}$ we have $\alpha(z + z') = \alpha(z) + \alpha(z')$, $\alpha(zz') = \alpha(z)\alpha(z')$ and $\alpha(q) = q$.

We now calculate the effect of applying α repeatedly to w. By $\alpha^2(w)$ we mean $\alpha(\alpha(w))$, and more generally, for integers $j > 1$, $\alpha^j(w)$ means $\alpha(\alpha^{j-1}(w))$. We have $\alpha(w) = w^3$ and $w^{17} = 1$. Also $\alpha(w^i) = (\alpha(w))^i$ for every positive integer i because α preserves products. Hence

$$\alpha(w) = w^3; \ \alpha^2(w) = \alpha(w^3) = (\alpha(w))^3 = (w^3)^3 = w^9;$$

$$\alpha^3(w) = \alpha(w^9) = (\alpha(w))^9 = w^{27} = w^{10};$$

and so on. In fact $\alpha^i(w) = w^j$ where $j \equiv 3^i \pmod{17}$. In particular, $\alpha^4(w) = w^{81} = w^{13} = 1/w^4$ and $\alpha^8(w) = w^{16} = 1/w$. Also $\alpha^{16}(w) = w$.

Let $K = \{x \in \mathbb{Q}(w) \mid \alpha^8(x) = x\}$. We leave it as an exercise for the reader to show that K is a subfield of $\mathbb{Q}(w)$. We have $\alpha^8(w) = 1/w \neq w$ so that $w \notin K$. Therefore $K \neq \mathbb{Q}(w)$ and so $[\mathbb{Q}(w):K] \neq 1$.

Let $L = \{x \in \mathbb{Q}(w) \mid \alpha^4(x) = x\}$. For each $x \in L$ we have $\alpha^8(x) = \alpha^4(\alpha^4(x)) = \alpha^4(x) = x$ so that $x \in K$. Thus $L \subseteq K$ and it is easily seen that L is a subfield of K. Set $r = w + \alpha^8(w) = w + 1/w$. Then

$$\alpha^8(r) = \alpha^8(w + 1/w) = \alpha^8(w + \alpha^8(w)) = \alpha^8(w) + \alpha^{16}(w) = 1/w + w = r.$$

Hence $r \in K$. However,

$$\alpha^4(r) = \alpha^4(w + 1/w) = \alpha^4(w) + \alpha^4(1/w)$$
$$= \alpha^4(w) + 1/\alpha^4(w) = w^{13} + w^4 \neq w + w^{16} = r$$

since $w, w^2, \ldots, w^{16}$ are linearly independent over $\mathbb{Q}$. Thus $r \in K$ but $r \notin L$, so that $[K:L] \neq 1$.

Let $M = \{x \in \mathbb{Q}(w) \mid \alpha^2(x) = x\}$. Once again it is easily seen that M is a subfield of L. Set $s = r + \alpha^4(r) = w + w^4 + w^{13} + w^{16}$. Then it can be checked that $\alpha^4(s) = s$ but $\alpha^2(s) \neq s$. Hence $s \in L$ but $s \notin M$. Therefore $[L:M] \neq 1$. Set

$$t = s + \alpha^2(s) = w + w^4 + w^{13} + w^{16} + w^9 + w^2 + w^{15} + w^8.$$

Then $\alpha^2(t) = t$ and so $t \in M$. However, $t \notin \mathbb{Q}$. Hence $M \neq \mathbb{Q}$ and so $[M:\mathbb{Q}] \neq 1$.

Now by 9.24 we have $16 = [\mathbb{Q}(w):\mathbb{Q}] = [\mathbb{Q}(w):K][K:L][L:M][M:\mathbb{Q}]$. We have shown above that none of the terms on the right-hand side is 1. Hence they must all be 2. By Question 4, Exercises 11, we have $\mathbb{Q}(w) = K(\sqrt{k})$, $K = L(\sqrt{l})$, $L = M(\sqrt{m})$ and $M = \mathbb{Q}(\sqrt{q})$ for some $k \in K$, $l \in L$, $m \in M$ and $q \in \mathbb{Q}$. This is enough to show that the regular 17-gon is constructible.

The above, however, does not give a method for constructing a 17-gon because we do not know the values of k, l, m, and q. Since $r = w + 1/w$ we have $w^2 - rw + 1 = 0$. Therefore $w = (r + \sqrt{(r^2 - 4)})/2$ for some choice of the square root. Since $s = r + \alpha^4(r)$ we have $r^2 - sr + r\alpha^4(r) = 0$. Direct calculation gives

$$r\alpha^4(r) = (w + w^{16})(w^4 + w^{13})$$
$$= w^5 + w^{14} + w^3 + w^{12} = -s^3/2 + 3s - 3/2.$$

The last equality can be checked by substituting for s in terms of w. Now the equation $r^2 - sr + r\alpha^4(r) = 0$ gives

$$2r^2 - 2sr - s^3 + 6s - 3 = 0.$$

Hence

$$r = (s + \sqrt{(2s^3 + s^2 - 12s + 6)})/2.$$

Similar but easier calculations give

$$0 = s^2 - ts + s\alpha^2(s) = s^2 - ts - 1.$$

so that $s = (t + \sqrt{(t^2 + 4)})/2$. Also

$$0 = t^2 - (t + \alpha(t))t + t\alpha(t) = t^2 + t - 4$$

and hence $t = (\sqrt{17} - 1)/2$ for some choice of the square root in each case. This enables us to write any element of $\mathbb{Q}(w)$ explicitly in terms of square roots. ∎

Comments 12.26.   Roughly speaking a real number is constructible if and only if it can be obtained from rational numbers by a finite number of extractions of square roots. The regular 17-gon is constructible because we can express $\cos(360/17)$ in this form. We have obtained: $\cos(360/17) = r/2$, where

$$r = (s + \sqrt{(2s^3 + s^2 - 12s + 6)})/2,$$

where $s = (t + \sqrt{(t^2 + 4)})/2$, where $t = (\sqrt{17} - 1)/2$ for some choice of square roots. This gives a precise but complicated procedure for constructing $\cos(360/17)$. For a quicker construction see Coxeter (1961, p. 27) or Rouse Ball and Coxeter (1974, p. 45).

Exercises 12

1. Show how to construct an equilateral triangle.

2. Show that $641 \mid 2^{32} + 1$.

3. Verify that 3 is a generator for the cyclic group of non-zero elements of $\mathbb{Z}/17\mathbb{Z}$ under multiplication.

4. Let A and B be distinct points in the plane. Show how to construct a line through B at right angles to AB.

5. Let p be a prime number with $p = 2^k + 1$ for some positive integer k. Show that k is a power of 2, i.e. that p is a Fermat prime.

6. Show that the regular 2^k-gon is constructible for every positive integer k (the case $k = 1$ being considered trivial).

7. Show that $w, w^2, \ldots, w^{16}$ form a basis for $\mathbb{Q}(w)$ as a vector space over $\mathbb{Q}$ in 12.25.

8. Show that K is a subfield of $\mathbb{Q}(w)$ in 12.25.

9. Let n be a positive integer with $n \neq 1$ and $n \neq 2$. Show that the regular n-gon is constructible if and only if $\varphi(n)$ is a power of 2.

10. Set $c = 2\cos(72) = w + 1/w$ where $w = e^{2i\pi/5}$. Use the fact that $w^5 - 1 = 0$ to show that $w^4 + w^3 + w^2 + w + 1 = 0$. Hence or otherwise show that $c^2 + c - 1 = 0$ and that $\cos(72) = (\sqrt{5} - 1)/4$.

13 Homomorphisms, ideals, and factor rings

It is often the case that two rings for which we already have some information are in fact related to each other. For instance the ring $\mathbb{Z}/n\mathbb{Z}$ of integers mod n is in some sense derived from the ring $\mathbb{Z}$ of integers. We shall now study such connections between rings. This is done by means of homomorphisms, which are the appropriate type of function from one ring to another. One reason for doing this is that it helps us to organize our examples of rings by establishing natural connections between some of them.

Closely related to homomorphisms are ideals which are certain subsets of a ring. The properties of the two are intimately connected. This theory then allows us to construct new examples of rings with certain specified properties. We shall give an important application of this type in Chapters 14 and 15, where we shall show how to construct a finite field with p^n elements for any choice of the prime number p and positive integer n.

Definition 13.1. Let R and S be rings. A *homomorphism* from R to S is a function $f: R \rightarrow S$ such that (i) $f(a+b) = f(a) + f(b)$ and (ii) $f(ab) = f(a)f(b)$ for all $a, b \in R$.

Some authors also require that $f(1_R) = 1_S$ where 1_R and 1_S are the identity elements of R and S respectively. However, we do not include this as part of the definition.

We can immediately deduce an important property of homomorphisms.

Proposition 13.2. *Let R and S be rings and let $f: R \rightarrow S$ be a homomorphism. Then*

(i) $f(0_R) = 0_S$ *where 0_R and 0_S are the zero elements of R and S respectively.*

(ii) $f(-a) = -f(a)$ *for all $a \in R$.*

Proof. (i) We have

$$f(0_R) = f(0_R + 0_R)$$
$$= f(0_R) + f(0_R),$$

since f is a homomorphism. Now S contains the element $-f(0_R)$. Adding it to the two sides of the equation we have $0_S = f(0_R)$.

(ii) We have

$$0_S = f(0_R) = f(a + (-a))$$
$$= f(a) + f(-a),$$

since f is a homomorphism. By the uniqueness of $-f(a)$ in the additive group S we must have $f(-a) = -f(a)$.   ■

Example 13.3. Let n be a fixed positive integer and let f be the function which sends each integer x to the corresponding integer $x \pmod{n}$. Then f is a homomorphism from $\mathbb{Z}$ to $\mathbb{Z}/n\mathbb{Z}$.   ■

Example 13.4. Let $S = \mathbb{R}[X]$. We define $f: S \to \mathbb{C}$ as follows: Let $s \in S$. Then $s = g(X)$ for some polynomial $g(X)$ with coefficients in $\mathbb{R}$. Set $f(s) = g(i)$, i.e. write s in terms of X and then replace X by i. Then f is a homomorphism from S to $\mathbb{C}$.   ■

Example 13.5. Let R be a ring. Define $f: R \to R$ by $f(x) = x$ for all $x \in R$; thus f is the identity function on R. Then f is a homomorphism. ■

A homomorphism $f: R \to S$ from a ring R to a ring S does not automatically map the identity of R onto the identity of S, as the next example shows.

Example 13.6. Let S be the ring of all pairs of integers with operations given by $(a, b) + (c, d) = (a + c, b + d)$ and $(a, b)(c, d) = (ac, bd)$. Define $f: \mathbb{Z} \to S$ by $f(x) = (x, 0)$ for all $x \in \mathbb{Z}$. Then f is a homomorphism but $f(1) = (1, 0)$ is not the identity of S.   ■

We now recall some elementary definitions. Let A, B be non-empty sets and $f: A \to B$ a function.

(i) f is said to be *surjective* (or onto) if given $b \in B$ there exists at least one $a \in A$ such that $b = f(a)$.

(ii) f is said to be *injective* (or one to one) if $f(a_1) = f(a_2)$ implies that $a_1 = a_2$ for $a_1, a_2 \in A$.

(iii) f is called a *bijection* (or a bijective function) if it is both injective and surjective.

Let $f: A \to B$ be a bijection. In this case the *inverse* function $f^{-1}: B \to A$ can be defined. We define for $b \in B$, $f^{-1}(b) = a$ where a is the unique element such that $f(a) = b$.

Note that the symbol $f^{-1}(b)$ is an element of the set A. It does *not* mean $1/f(b)$, which in any case may not be defined.

Definition 13.7. Let R and S be rings. A bijective homomorphism from R onto S is called an *isomorphism*. We say that S is isomorphic to R if there is at least one isomorphism from R to S. An isomorphism from R to R is called an *automorphism*.

Note that by Question 1, Exercises 13, if f is an isomorphism from R onto S then f^{-1} is an isomorphism from S onto R. Because of this symmetry we are justified in saying that R and S are isomorphic rings when there is an isomorphism from R onto S. The symbol $R \cong S$ denotes the fact that R and S are isomorphic rings.

If R and S are isomorphic rings, they have identical properties as *rings*. Thus let f be an isomorphism between the two rings and let $r \in R$. Then, for example, $r^n = 0$ implies that $f(r)^n = 0$; r is a unit of R implies that $f(r)$ is a unit of S and if R is a field then so also is S. (See Question 4, Exercises 13.)

Example 13.8. The function f given in 13.3 is not an isomorphism because it is not injective. We have, for example, $f(n) = f(0) = 0$. In fact $\mathbb{Z}$ and $\mathbb{Z}/n\mathbb{Z}$ are not isomorphic because $\mathbb{Z}$ is infinite but $\mathbb{Z}/n\mathbb{Z}$ is finite. ∎

Example 13.9. In 13.4 we have $f(X^2 + 1) = i^2 + 1 = 0 = f(0)$. Hence f is not injective and so f is not an isomorphism. In fact S and $\mathbb{C}$ are not isomorphic because $\mathbb{C}$ is a field and S is not. ∎

Example 13.10. Let f be the identity function on a ring R as in 13.5. Then f is an automorphism of R. ∎

Example 13.11. The homomorphism $f: \mathbb{Z} \to S$ in 13.6 is not an isomorphism because it is not surjective. In fact $\mathbb{Z}$ and S are not isomorphic because $\mathbb{Z}$ is an integral domain while S is not. ∎

Example 13.12. Let $f: \mathbb{C} \to \mathbb{C}$ be the function $f(x + iy) = x - iy$ for all $x, y \in \mathbb{R}$. Then f is an automorphism of $\mathbb{C}$. ∎

The injective property for a homomorphism can usually be easily checked using the next criterion.

Proposition 13.13. *Let R, S be rings and $f: R \to S$ a homomorphism. Then f is injective if and only if 'for $x \in R$, $f(x) = 0$ implies that $x = 0$'.*

Proof. Suppose that f is injective and that $f(x) = 0$ for some $x \in R$. By 13.2 $f(0) = 0$. Thus $f(x) = f(0)$ and so $x = 0$ since f is injective.

Conversely, suppose that the given condition holds. Let $a, b \in R$ with $f(a) = f(b)$. Then $f(a - b) = f(a) - f(b) = 0$. By the assumption on f we have $a - b = 0$ and so $a = b$. Thus f is injective. ∎

Example 13.14 (The Frobenius automorphism). Let F be a finite field with p^n elements for some prime p and positive integer n. Let $f : F \to F$ be the function given by $f(a) = a^p$ for all $a \in F$. Then for all $a, b \in F$ we have $f(ab) = (ab)^p = a^p b^p = f(a)f(b)$. We showed earlier in Question 8, Exercises 9 that $(a + b)^p = a^p + b^p$. Hence $f(a + b) = f(a) + f(b)$. Thus f is a homomorphism.

We shall now check that f is both injective and surjective. Let $a \in F$ with $f(a) = 0$. Then $a^p = 0$ and so $a = 0$. So f is injective. As f is an injective function from a finite set into itself it must also be surjective. Thus f is an automorphism of F. It is known as the *Frobenius automorphism*. Since f is surjective we know that every element of F is expressible in the form a^p for some $a \in F$. ∎

Proposition 13.15. *The identity function is the only automorphism of $\mathbb{Z}$.*

Proof. Let f be an automorphism of $\mathbb{Z}$ and set $a = f(1)$. Then $a^2 = f(1)f(1) = f(1 \cdot 1) = f(1) = a$. Hence $a(a - 1) = 0$. Therefore either $a = 0$ or $a = 1$. Since f is injective we must have $f(1) \ne f(0)$ so that $a \ne 0$. Hence $a = 1$, i.e. $f(1) = 1$. Hence $f(2) = f(1 + 1) = f(1) + f(1) = 1 + 1 = 2$; $f(3) = f(2 + 1) = f(2) + f(1) = 2 + 1 = 3$ and so on. Thus $f(n) = n$ for every positive integer n. Now let k be a negative integer. Then $-k$ is positive. Therefore by the above, $f(-k) = -k$. Hence $-f(k) = -k$ and so $f(k) = k$. Since we also have $f(0) = 0$, we have shown that $f(x) = x$ for all $x \in \mathbb{Z}$. Thus f is the identity function. ∎

Definition 13.16. Let f be a homomorphism from a ring R to a ring S. The set $\{r \in R : f(r) = 0\}$ is called the *kernel* of f. It is denoted by $\ker f$.

Proposition 13.17. *Let f be a homomorphism from a ring R to a ring S. Set $K = \ker f$. Then*

(i) *$0 \in K$,*

(ii) *If $a, b \in K$ then $a - b \in K$,*

(iii) *If $a \in K$ and $r \in R$ then $ar \in K$.*

Proof. (i) We have proved this in 13.2.

(ii) Let $a, b \in K$. Then $f(a) = 0$ and $f(b) = 0$. Therefore $f(a - b) = f(a) - f(b) = 0$ so that $a - b \in K$.

(iii) Let $a \in K$ and $r \in R$. Then $f(a) = 0$ and so $f(ar) = f(a)f(r) = 0 \cdot f(r) = 0$. Hence $ar \in K$.   ∎

Definition 13.18. A subset K of a ring R is called an *ideal* of R if K satisfies (i), (ii) and (iii) of 13.17.

Comments 13.19

(a) We showed in 13.17 that the kernel of a homomorphism is an ideal. We shall show in 13.37 that if R is a ring and K is an ideal of R then there is a homomorphism f from R to some ring S such that $\ker f = K$. Thus a nonempty subset of R is an ideal if and only if it is the kernel of a homomorphism from R to some ring S.

(b) Let R be a ring. Then R is an ideal of R. Also the subset $\{0\}$ consisting of the element 0 alone is an ideal of R.

(c) Let K be an ideal of a ring R. If $K = R$ then clearly $1 \in K$. Conversely, let $1 \in K$. Then for $r \in R$ we have $r = 1 \cdot r \in K$. Hence $K = R$. Thus $1 \in K$ if and only if $K = R$.

(d) In some respects, ideals of rings correspond to normal subgroups of groups. This is so if we think in terms of kernels of homomorphisms. However, the analogy must not be pushed too far. For example, if N is a normal subgroup of a group G then it is a subgroup of G. On the other hand the set $2\mathbb{Z}$ of even integers is an ideal of the ring $\mathbb{Z}$ but it is *not* a subring according to our definitions. Also every subgroup of an Abelian group is a normal subgroup, but not every subring of a commutative ring is an ideal. For example, $\mathbb{Z}$ is a subring of $\mathbb{Q}$. However, $2 \in \mathbb{Z}$ and $1/3 \in \mathbb{Q}$, yet $2 \cdot (1/3) \notin \mathbb{Z}$ so that $\mathbb{Z}$ is not an ideal of $\mathbb{Q}$. Alternatively, $1 \in \mathbb{Z}$ but $\mathbb{Z} \neq \mathbb{Q}$. So by (c) $\mathbb{Z}$ cannot be an ideal of $\mathbb{R}$.

Example 13.20. Let n be a fixed integer and let $n\mathbb{Z}$ be the set of all multiples of n. Clearly $0 = n \cdot 0 \in n\mathbb{Z}$. Let $a, b \in n\mathbb{Z}$. Then $a = nc$ and $b = nd$ for some $c, d \in \mathbb{Z}$. So $a - b = n(c - d) \in n\mathbb{Z}$. Similarly $a \in n\mathbb{Z}$ and $r \in \mathbb{Z}$ implies that $ar \in n\mathbb{Z}$. Thus $n\mathbb{Z}$ is an ideal of $\mathbb{Z}$.

Conversely, let K be an ideal of $\mathbb{Z}$. Then K is, in particular, an additive subgroup of the cyclic group $\mathbb{Z}$. Hence K is cyclic as an additive group. Thus there is an integer n such that K consists of all the multiples of n, i.e. $K = n\mathbb{Z}$.

Thus a non-empty subset of $\mathbb{Z}$ is an ideal if and only if it is of the form $n\mathbb{Z}$ for some $n \in \mathbb{Z}$.   ∎

Example 13.21. Let R be a ring and let x be a fixed element of R. Let $K = \{r \in R \mid xr = 0\}$. We have $x \cdot 0 = 0$ so that $0 \in K$. Let $a, b \in K$. Then $xa = 0$ and $xb = 0$ so that $x(a - b) = xa - xb = 0$. Hence $a - b \in K$. With $a \in K$ and $r \in R$ we have $x(ar) = (xa)r = 0 \cdot r = 0$ and so $ar \in K$. Therefore K is an ideal of R.   ∎

Definition 13.22. Let K be an ideal of the ring R and let $a \in R$. The set $a + K = \{a + k \mid k \in K\}$ is called a *coset* (of a with respect to K).

Example 13.23. Let $R = \mathbb{Z}/6\mathbb{Z}$ and let K be the set $\{0, 3\}$. Then it is easily seen that K is an ideal of R. We shall determine all the cosets in R with respect to K. Firstly $0 + K = K = \{0, 3\}$ is a coset. Next we try $1 + K$. We have $1 + K = \{1 + 0, 1 + 3\} = \{1, 4\}$. We now try $2 + K$. We see that $2 + K = \{2 + 0, 2 + 3\} = \{2, 5\}$. These are all the cosets of K in R. The cosets $3 + K$, $4 + K$, $5 + K$ do not give anything new. For example, $4 + K = \{4 + 0, 4 + 3\} = \{1, 4\} = 1 + K$.   ∎

Further properties of cosets are described in 13.24 and 13.25 below.

As the above example shows, the crucial point when dealing with cosets is that it is possible to have $a + K = b + K$ with $a \neq b$. The next proposition shows exactly when this happens.

Proposition 13.24. *Let R be a ring and K an ideal of R. Let $a, b \in R$. Then $a + K = b + K \Leftrightarrow a - b \in K$.*

Proof.

$$a - b \in K \Rightarrow a - b = k_1 \qquad \text{for some } k_1 \in K$$
$$\Rightarrow a + k = b + k + k_1 \qquad \text{for all } k \in K$$
$$\Rightarrow a + K \subseteq b + K.$$

Similarly $a - b \in K \Rightarrow b + K \subseteq a + K$ and hence

$$a - b \in K \Rightarrow a + K = b + K.$$

Conversely, $a + K = b + K \Rightarrow a \in b + K \Rightarrow a = b + k$ for some $k \in K \Rightarrow a - b \in K$.   ∎

Comments 13.25.   Let R be a ring.

(a) We can form cosets $a + K$ as in 13.22 when K is any additive subgroup of R. However, in the context of rings, cosets play a useful role only when K is an ideal of R.

(b) Let K be an ideal of a ring R and $a, b \in R$. Under addition, K is a subgroup of the group R. The coset $a + K$ defined above is the same as that defined in group theory. Thus 13.24 corresponds to the following familiar fact: if H is a subgroup of a group G then, using the usual multiplicative notation, $Hx = Hy$ if and only if $xy^{-1} \in H$ for $x, y \in G$.

   Further properties which are easily proved and which we quote from group theory are:

   (i) $a + K$ and $b + K$ have the same number of elements (i.e. there is a bijection between the two sets).

   (ii) Either $a + K = b + K$ or $(a + K) \cap (b + K) = \varnothing$. Each element of R belongs to one and only one coset of K in R.

(c) Note that $a = a + 0 \in a + K$.

Example 13.26.   Let n be a fixed integer. Set $K = n\mathbb{Z}$, i.e. K is the set of all integers which are divisible by n. As shown in 13.20, $n\mathbb{Z}$ is an ideal of $\mathbb{Z}$. We fix $a \in \mathbb{Z}$. Then $a + K = \{a + k \mid k \in K\} = \{a + k \mid k \in n\mathbb{Z}\} = \{a + nz \mid z \in \mathbb{Z}\}$. Thus $a + n\mathbb{Z}$ consists of all integers which have the same remainder as a when divided by n. Therefore the cosets of $n\mathbb{Z}$ in $\mathbb{Z}$ are just the congruence classes (mod n). ∎

Definition 13.27 (Addition and multiplication of cosets).   Let K be an ideal of a ring R and let $a, b \in R$. We define the sum of the cosets $a + K$ and $b + K$ to be the coset $a + b + K$, and the product of $a + K$ and $b + K$ to be the coset $ab + K$.

   Of course, since each coset can be represented in more than one way, we must immediately verify the consistency of these definitions.

Proposition 13.28.   *Let K be an ideal of a ring R. Then the addition and multiplication of cosets stated in 13.27 are well defined.*

Proof.   Let $a, b, c, d \in R$. Suppose that $a + K = c + K$ and $b + K = d + K$. We are required to show that (i) $a + b + K = c + d + K$ and (ii) $ab + K = cd + K$.

   By 13.24 we have $a - c \in K$ and $b - d \in K$. Hence $(a + b) - (c + d) \in K$ and so $a + b + K = c + d + K$ by 13.24 proving (i).

   Similarly, $ab - cd = a(b - d) + (a - c)d \in K$ since K is an ideal of R. So again by 13.24 we have $ab + K = cd + K$, proving (ii). ∎

Comments 13.29. In 13.28 the proof that the multiplication of cosets with respect to K is well defined requires the fact that $rk \in K$ whenever $r \in R$ and $k \in K$. The next example shows what can go wrong with the multiplication of cosets if K is a subring rather than an ideal of R.

Non-example 13.30. Take $R = \mathbb{Q}$ and $K = \mathbb{Z}$. Let $a, b \in \mathbb{Q}$. By 13.24 we have $a + \mathbb{Z} = b + \mathbb{Z}$ if and only if $a - b \in \mathbb{Z}$. Thus $2/7 + \mathbb{Z} = 16/7 + \mathbb{Z}$ and $4/3 + \mathbb{Z} = 1/3 + \mathbb{Z}$. However, working with the product given in 13.27 we obtain

$$(2/7 + \mathbb{Z})(4/3 + \mathbb{Z}) = 8/21 + \mathbb{Z} \neq 16/21 + \mathbb{Z} = (16/7 + \mathbb{Z})(1/3 + \mathbb{Z}),$$

showing that the product is not well-defined in this case.   ∎

Example 13.31. As in 13.23 set $R = \mathbb{Z}/6\mathbb{Z}$ and let K be the ideal of R consisting of the two elements 0 and 3. We shall check the consistency of coset addition and multiplication on this example. We introduce the following notation for the three cosets of K in $R: O = \{0, 3\}$, $I = \{1, 4\}$ and $J = \{2, 5\}$. We can calculate $I + J$ by choosing any $x \in I$ and any $y \in J$, forming $x + y$ in R, and finding the coset which contains $x + y$. For example, we can take $x = 1$ and $y = 5$, so that $x + y = 6 = 0$ which is in O. Hence $I + J = O$. Alternatively, we could have chosen $x = 4$ and $y = 5$ so that $x + y = 9 = 3$ which is again in O. In fact $x + y \in O$ for all $x \in I$ and all $y \in J$. This is just a particular instance of addition being well defined, as was shown in 13.28. In order to calculate IJ we choose $x = 4$ from I and $y = 2$ from J, so that $xy = 8 = 2$ which is in J. Hence $IJ = J$. To find J^2 we choose $y = 2$ from J, so that $y^2 = 4$ which is in I. Hence $J^2 = I$; and so on.

It is easily seen that the set of cosets O, I, J with coset operations is a ring which is isomorphic to $\mathbb{Z}/3\mathbb{Z}$ where O, I, J correspond to 0, 1, 2 respectively.   ∎

Example 13.32. Let n be a fixed positive integer and set $K = n\mathbb{Z}$. Then there are n distinct cosets of K in $\mathbb{Z}$, namely

$$K = 0 + K, 1 + K, 2 + K, \ldots, n - 1 + K.$$

Addition and multiplication of these cosets correspond exactly to the addition and multiplication in $\mathbb{Z}/n\mathbb{Z}$.   ∎

Proposition 13.33. *Let K be an ideal of the ring R. Then the set of cosets of K in R is a ring with respect to the addition and multiplication of cosets defined in 13.27.*

Proof. This involves routine checking. Let us, for example, establish the associative law for multiplication. Let $a, b, c \in R$. Then

$$((a+K)(b+K))(c+K) = (ab+K)(c+K) = abc + K$$
$$= (a+K)(bc+K) = (a+K)((b+K)(c+K)).$$

The other axioms for a ring can be verified similarly. The zero element is the coset $0 + K$, the multiplicative identity is $1 + K$, and $-(a+K) = -a + K$. ∎

Definition 13.34. Let K be an ideal of a ring R. The set of cosets of K in R with addition and multiplication as in 13.27 is called the *factor ring* of R by K and is denoted by R/K. Some authors call R/K a *quotient ring* of R.

Comments 13.35. Let K be an ideal of a ring R. Roughly speaking, we form the ring R/K by putting all the elements of K equal to 0, and as a consequence we identify elements a, b of R if $a - b \in K$ (see 13.24). With R and K as in 13.31 the formation of R/K can be thought of as the following procedure: 'Take integers (mod 6) and force $0 = 0$ and $3 = 0$'. It is then not surprising that we end up with integers (mod 3). With $R = \mathbb{Z}$ and $K = n\mathbb{Z}$ as in 13.32 the formation of R/K can be thought of as 'taking $\mathbb{Z}$ and equating all multiples of n to 0'. We end up with the integers (mod n). This explains the use of the symbol $\mathbb{Z}/n\mathbb{Z}$ for the integers (mod n). Indeed, formally the most satisfactory way of defining the ring of integers (mod n) is as the factor ring $\mathbb{Z}/n\mathbb{Z}$. From a practical computational point of view, however, cosets do not provide us with easy and useful notation.

Proposition 13.36. *Let S and T be rings and let R be the ring*

$$\{(s,t) \mid s \in S, t \in T\}$$

with the usual componentwise operations. Let $K = \{(0,t) \mid t \in T\}$. Then K is an ideal of R and $R/K \cong S$.

Proof. We leave it as an exercise to show that K is an ideal of R.

A typical element of R/K is the coset $(s,t) + K$ where $s \in S$ and $t \in T$. Define $f: R/K \to S$ by $f((s,t)+K) = s$. We must first check that f is well defined. Let $s, s' \in S$ and $t, t' \in T$. Suppose that $(s,t) + K = (s',t') + K$. We must show that $s = s'$. By 13.24 we have $(s,t) - (s',t') \in K$ and hence $(s-s', t-t') \in K$. Therefore $s - s' = 0$. Thus $s = s'$ as required. It is now routine to check that f is an isomorphism. ∎

Proposition 13.37. *Let* K *be an ideal of the ring* R. *Then there is a homomorphism* $f: R \to R/K$ *such that* $\ker f = K$.

Proof. Set $f(r) = r + K$ for all $r \in R$. Then for all $a, b \in R$ we have

$$f(a + b) = a + b + K = (a + K) + (b + K) = f(a) + f(b)$$

and

$$f(ab) = ab + K = (a + K)(b + K) = f(a)f(b).$$

Thus f is a homomorphism. We have $f(a) = 0 \Leftrightarrow a + K$ is the zero coset $\Leftrightarrow a + K = 0 + K \Leftrightarrow a \in K$ by 13.24. Hence $\ker f = K$. ∎

Comments 13.38. The homomorphism f defined in the proof of 13.37 is sometimes called the *canonical epimorphism* from R to R/K. It is a matter of personal taste as to whether one prefers to use the coset notation $a + K$ or the function notation $f(a)$. In the function notation we have:

  (i) every element of R/K is of the form $f(a)$ for some $a \in R$;

  (ii) $f(a + b) = f(a) + f(b)$ and $f(ab) = f(a)f(b)$;

  (iii) $f(a) = 0$ if and only if $a \in K$;

  (iv) $f(1)$ is the identity element of the ring R/K.

Definition 13.39. Let M be an ideal of a ring R. Then M is called a *maximal ideal* of R if (i) $M \neq R$ and (ii) the only ideals of R which contain M are R and M.

Theorem 13.40. *Let* M *be an ideal of a ring* R. *Then* M *is a maximal ideal if and only if* R/M *is a field.*

Proof. We shall use the canonical epimorphism f from R to R/M as in 13.38. Thus f is a surjective homomorphism from R onto R/M with $\ker f = M$. Since M is always one of the cosets of M in R, the following statements are equivalent: (i) $M \neq R$; (ii) there are at least two cosets with respect to M in R; (iii) R/M has at least two elements; (iv) $1 \neq 0$ in R/M.

Suppose that M is a maximal ideal of R. Then $M \neq R$ so that in R/M we have $1 \neq 0$. Let $x \in R/M$, $x \neq 0$. We must show that $xy = 1$ for some $y \in R/M$. We have $x = f(a)$ for some $a \in R$. Since $x \neq 0$ in R/M, by 13.24 we have $a \notin M$. Let $I = \{ar + m \mid r \in R, m \in M\}$. We leave it as an exercise to show that I is an ideal of R. Now for $m \in M$ we have $m = a \cdot 0 + m$

with $0 \in R$. Hence $M \subseteq I$. However, $I \neq M$ since $a = a \cdot 1 + 0 \in I$ but $a \notin M$. As M is a maximal ideal of R we have $I = R$. In particular, $1 = ar + m$ for some $r \in R$ and $m \in M$. Hence $f(1) = f(ar + m) = f(a)f(r) + f(m) = f(a)f(r)$ because $M = \ker f$ and so $f(m) = 0$. Set $y = f(r)$. Then $xy = f(1)$ where $f(1)$ is the multiplicative identity element of R/M, i.e. $xy = 1$ in R/M.

Conversely suppose that R/M is a field. Then $1 \neq 0$ in R/M so that $M \neq R$. Let I be an ideal such that $I \supsetneq M$. We must show that $I = R$. Choose an element $a \in I$ such that $a \notin M$. Set $x = f(a)$. Then x is a non-zero element of the field R/M. Hence $xy = 1$ for some $y \in R/M$. We have $y = f(b)$ for some $b \in R$. Also $f(1) = 1$ in R/M. Thus $f(1) = xy = f(a)f(b) = f(ab)$. Hence $f(1 - ab) = f(1) - f(ab) = 0$ and so $1 - ab \in M$. Thus $1 - ab = m$ for some $m \in M$. Hence $1 = ab + m$. However, $m \in I$ since $M \subsetneq I$. Also $ab \in I$ since $a \in I$ and $b \in R$. Therefore $1 = ab + m \in I$. It follows by 13.19(c) that $I = R$. Thus M is a maximal ideal of R.   ∎

Comments 13.41. In Chapters 14 and 15 we shall apply Theorem 13.40 to construct finite fields in the form R/M for a suitable ring R and maximal ideal M of R. This demonstrates an important use of factor rings, namely to construct new rings with specified properties.

Theorem 13.42 (The fundamental homomorphism theorem). *Let R, S be rings and let f be a surjective homomorphism from R onto S. Then the ring R/K is isomorphic to S where $K = \ker f$.*

Proof. Let $g: R/K \to S$ be defined by $g(r + K) = f(r)$ for all $r \in R$. Since a coset of K in R can be written in more than one way, we must first check that g is well defined. Suppose that $a + K = b + K$ for some $a, b \in R$. We must show that $f(a) = f(b)$. Now by 13.24 we have $a - b \in K = \ker f$. So $f(a - b) = 0$ and $f(a) = f(b)$ as required.

For all $a, b \in R$ we have

$$g((a + K) + (b + K)) = g(a + b + K)$$
$$= f(a + b)$$
$$= f(a) + f(b) \quad \text{since } f \text{ is a homomorphism.}$$
$$= g(a + K) + g(b + K).$$

Similarly,

$$g((a + K)(b + K)) = g(ab + K)$$
$$= f(ab)$$
$$= f(a)f(b)$$
$$= g(a + K)g(b + K).$$

Hence g is a homomorphism.

Let $s \in S$. Since f is surjective, we have $s = f(a)$ for some $a \in R$. Hence $s = g(a + K)$. Therefore g is surjective. Now suppose that $g(r + K) = 0$ for some $r \in R$. Then $f(r) = 0$ and so $r \in \ker f = K$. Therefore $r + K = K$, i.e. $r + K$ is the zero element of R/K. By 13.13 g is injective and hence an isomorphism. ■

Comment 13.43. A standard use of the fundamental homomorphism theorem is as follows: Suppose that we are given a ring R and an ideal K of R and that we wish to determine the ring R/K. We may be told, or we may guess that there is a known ring S such that R/K is isomorphic to S. One way to justify this, if our guess is correct, is to find a surjective homomorphism from R to S such that $\ker f = K$. Then Theorem 13.42 shows that R/K is isomorphic to S. One way to remember 13.42 is that for a homomorphism f, 'the image of f is isomorphic to the domain of f factored by the kernel of f (in symbols, image $f \cong$ domain $f/\ker f$)'.

Example 13.44. As in Proposition 13.36 let S, T be rings and let $R = \{(s, t) \mid s \in S, t \in T\}$. Let $K = \{(0, t) \mid t \in T\}$. We can use the fundamental homomorphism theorem to show $R/K \cong S$ as follows. We must find a surjective homomorphism $f: R \to S$ so that $f((s, t)) \in S$; and we must arrange that $f((s, t)) = 0$ if and only if $s = 0$, so that $\ker f = K$. There is only one sensible way of defining f, namely $f((s, t)) = s$. It is then routine to check that f is a surjective homomorphism from R to S with $\ker f = K$. Therefore $R/K \cong S$ by 13.42. ■

Example 13.45. Let S be any ring and set $R = S[X]$. Let K be the set of all polynomials in R which have zero constant term. It is easily shown that K is an ideal of R. The elements of K are of the form $X \cdot f(X)$ for some $f(X) \in R$. Thus forming R/K seems to mean 'put $X = 0$'. We guess that $R/K \cong S$. To justify this define $h: R \to S$ as follows. Let $r \in R$. Then $r = s_0 + s_1 X + \cdots + s_n X^n$ for some non-negative integer n and some $s_i \in S$. Set $h(r) = s_0$. It is routine to check that h is a surjective homomorphism with $\ker h = K$. Therefore $R/K \cong S$ by 13.42. ■

Exercises 13

1. Let f be an isomorphism from a ring R to a ring S. Show that f^{-1} is an isomorphism from S onto R.

2. Let $f: R \to S$ and $g: S \to T$ be isomorphisms where R, S, and T are rings. Show that gf is an isomorphism from R onto T, where $gf(r) = g(f(r))$ for all $r \in R$.

3. Let R be a ring. Prove that the set of automorphisms of R is a group with respect to composition of functions.

4. Let R and S be isomorphic rings. Prove that R is a field if and only if S is a field, and that R is an integral domain if and only if S is an integral domain.

5. Let the function $f: \mathbb{Q}(\sqrt{2}) \to \mathbb{Q}(\sqrt{2})$ be defined by $f(a + b\sqrt{2}) = a - b\sqrt{2}$ for all $a, b \in \mathbb{Q}$. Prove that f is an automorphism of $\mathbb{Q}(\sqrt{2})$.

6. Let F be a field. Prove that the only ideals of F are F and $\{0\}$, the ideal consisting of the element 0 only.

7. Let M be an ideal of $\mathbb{Z}$. Prove that M is a maximal ideal of $\mathbb{Z}$ if and only if $M = p\mathbb{Z}$ for some prime number p.

8. Let R be the ring of all polynomials in X with coefficients in $\mathbb{Z}$, and let K be the ideal of R consisting of all elements of R with an even constant term. Find a familiar ring which is isomorphic to R/K and justify your answer.

9. Let R be the ring of all pairs of integers with componentwise operations. Let $K = \{(a, b) \mid a, b \text{ even integers}\}$. Prove that K is an ideal of R and find the cosets of K in R.

   Determine the ring R/K either by deriving its addition and multiplication tables or by showing that it is isomorphic to some familiar ring.

10. Fill in the details in the following proof that the identity function is the only automorphism on $\mathbb{R}$.

    Let f be an automorphism of $\mathbb{R}$. As in the proof of 13.15 we have $f(z) = z$ for all $z \in \mathbb{Z}$. Let $a, b \in \mathbb{Z}$ with $b \neq 0$. Then $f(a/b) = f(a)/f(b) = a/b$. Hence f acts as the identity function on $\mathbb{Q}$. Let x be a positive real number. Then $x = y^2$ for some $y \in \mathbb{R}$. Then $f(x) = f(y^2) = (f(y))^2 > 0$. Hence f sends positive elements of $\mathbb{R}$ to positive elements of $\mathbb{R}$. It follows that if $u, v \in \mathbb{R}$ with $u < v$ then $f(u) < f(v)$. Let k be a fixed real number. For every $a \in \mathbb{Q}$ with $a < k$ we have $a = f(a) < f(k)$. Similarly for every $b \in \mathbb{Q}$ with $b > k$ we have $b > f(k)$. Therefore $f(k) = k$ for all $k \in \mathbb{R}$.

11. Let F be a finite field with p^n elements for some prime p and positive integer n, and let G be the automorphism group of F. Fill in the details of the following proof that G is cyclic of order n and is generated by the Frobenius automorphism f defined in 13.14. Set $q = p^n$. There is an element $t \in F$ such that the elements of F are $0, 1, t, t^2, \ldots, t^{q-2}$. We have $f(t) = t^p \neq t$ if $n \neq 1$, so that f is not the identity function. Also $f^2(t) = f(f(t)) = f(t^p) = (f(t))^p = (t^p)^p \neq t$ if

$n \neq 2$, so that f^2 is not the identity function. Continuing in this way we find that f^r is not the identity function for any integer r with $1 \leq r \leq n-1$. However, $f^n(t) = t^q = t$, from which it follows that $f^n(x) = x$ for all $x \in F$. Therefore f has order n as an element of G. Let S be the subfield of F with p elements. As in Question 11, Exercises 11, we have $F = S(t)$ and t is algebraic of degree n over S. Let $m(x)$ be the minimum polynomial of t over S. Then $\partial(m) = n$. By Question 9, Exercises 9 we have $m(X^p) = (m(X))^p$. Hence $m(f(t)) = m(t^p) = (m(t))^p = 0$. It follows that $t, f(t), f^2(t), \ldots, f^{n-1}(t)$ are the roots in F of the equation $m(X) = 0$. Let α be any automorphism of F. Then $\alpha(0) = 0$ and $\alpha(1) = 1$. It follows that $\alpha(s) = s$ for all $s \in S$. We now claim that $m(\alpha(t)) = \alpha(m(t))$. Let

$$m(X) = s_0 + s_1 X + \cdots + s_n X^n$$

for some $s_i \in S$. As $\alpha(s_i) = s_i$ we have

$$m(\alpha(t)) = s_0 + s_1 \alpha(t) + \cdots + s_n(\alpha(t))^n$$
$$= \alpha(s_0) + \alpha(s_1)\alpha(t) + \cdots + \alpha(s_n)\alpha(t)^n$$
$$= \alpha(s_0 + s_1 t + \cdots + s_n t^n)$$
$$= \alpha(m(t))$$

Thus $m(\alpha(t)) = \alpha(m(t)) = 0$ and so $\alpha(t) = f^j(t)$ for some j. It follows that $\alpha = f^j$. Therefore f generates G.

14 Principal ideal domains and a method for constructing fields

The simplest way to construct an ideal I of a ring R is to take I to be the set of elements of R which are multiples of (or equivalently, are divisible by) some fixed element $x \in R$. Some rings R have the special property that every ideal I of R arises in this way, i.e. $I = xR$ for some $x \in R$. Integral domains with this property are known as principal ideal domains. Examples of such rings include $\mathbb{Z}$ the ring of integers, J the ring of Gaussian integers, and the polynomial ring $F[X]$ in one indeterminate X over an arbitrary field F. In fact, every Euclidean domain is a principal ideal domain, and every principal ideal domain is a unique factorization domain. The converse of each of these statements is false.

In this chapter we shall devote our attention to principal ideal domains. As a particular application of this study, we shall construct finite fields as factor rings of $F[X]$, where F is the field $\mathbb{Z}/p\mathbb{Z}$ (p prime).

Definition 14.1. Let R be a ring and $a \in R$. Then the ideal $aR = \{ar \mid r \in R\}$ is called the *principal ideal* generated by a. Thus aR is the set of elements which are divisible by a.

An ideal I of a ring R is said to be *principal* if $I = aR$ for some $a \in R$.

Comments 14.2

(a) We already know (13.20) that every ideal of $\mathbb{Z}$ is principal. We shall show in 14.5 that every Euclidean domain has the same property.

(b) Let $a \in R$ where R is a ring. Then $a \in aR$ because $a = a \cdot 1$ with $1 \in R$. Let I be any ideal of R such that $a \in I$. Then $ar \in I$ for all $r \in R$. Hence $I \supseteq aR$.

(c) It is possible to have two elements $a, b \in R$ such that $aR = bR$ but $a \neq b$. In fact if R is an integral domain then $aR = bR$ if and only if $a = bu$ for some unit $u \in R$ (Question 2, Exercises 14). In particular, for $u \in R$, $uR = R$ if and only if u is a unit of R.

Non-example 14.3. Let $R = \mathbb{Z}[X]$. Let I be the ideal of R consisting of all elements of R which have even constant term. We shall show that I is not a principal ideal. Suppose to the contrary that $I = aR$ for some $a \in R$. Since $2 \in I$, we have $2 = ar$ for some $r \in R$. Hence a is a constant polynomial, i.e. a is an integer. Also $X \in I$ so that $X = as$ for some $s \in R$. We must now have either $a = 1$ and $s = X$ or $a = -1$ and $s = -X$. This is a contradiction because $a \in I$ and so a must be an even integer. ∎

Definition 14.4. An integral domain in which every ideal is principal is called a *principal ideal domain* (*PID*).

Theorem 14.5. *A Euclidean domain is a PID.*

Proof. Let R be an integral domain which is Euclidean with respect to the function N and let I be an ideal of R. If $I = \{0\}$ then $I = aR$ with $a = 0$. So assume that I contains at least one non-zero element. Let a be a non-zero element of I such that the non-negative integer $N(a)$ is as small as possible. Since $a \in I$ we have $aR \subseteq I$. Let $b \in I$. Since R is Euclidean we have $b = aq + r$ for some $q, r \in R$, where either $r = 0$ or $N(r) < N(a)$. Since I is an ideal and $a, b \in I$ we obtain $b - aq \in I$ and so $r \in I$. Suppose now that $r \neq 0$. Then we have $N(r) < N(a)$ with $r \neq 0$. This contradicts the choice of a. Therefore $r = 0$. Thus $b = aq$. Hence $I \subseteq aR$ and so $I = aR$. ∎

Corollary 14.6. *The ring of integers $\mathbb{Z}$, the ring of Gaussian integers J and the ring of polynomials $F[X]$, where F is a field, are principal ideal domains.*

Proof. We have shown that these rings are Euclidean in 2.8, 2.9 and 2.11. ∎

Summary 14.7

  (i) Every Euclidean domain is a PID (14.5).

 (ii) Not every PID is a Euclidean domain. The ring of all complex numbers of the form $(a + b\sqrt{(-19)})/2$ with $a, b \in \mathbb{Z}$ and $a + b$ even is such an example. The argument justifying this is beyond the scope of this book.

(iii) Every PID is a UFD (14.14).

(iv) Not every UFD is a PID. For example, $\mathbb{Z}[X]$ is a UFD (8.14(i)) but it is not a PID (14.3).

Our next major aim is to show that every PID is a UFD. We shall need to build up to this through several preliminary results.

Lemma 14.8. *Let R be a ring and $a, b \in R$. Then the following statements are equivalent*:

(i) $aR \subseteq bR$.

(ii) $a \in bR$.

(iii) $b \mid a$.

Proof. (ii) and (iii) are equivalent because they both mean that $a = br$ for some $r \in R$.

If $aR \subseteq bR$ then $a = a \cdot 1 \in aR$ and so $a \in bR$. Conversely, if $a \in bR$ then, since bR is an ideal, we have $aR \subseteq bR$. Hence (i) and (ii) are equivalent. Thus all three statements are equivalent. ∎

Lemma 14.9. *Let R be a PID and let $a_1, a_2, \ldots$ be a sequence of elements of R such that $a_1 R \subseteq a_2 R \subseteq a_3 R \subseteq \cdots$. Then there is a positive integer n such that $a_n R = a_{n+1} R = a_{n+2} R = \cdots$, i.e. $a_n R = a_s R$ for all $s \geq n$.*

Proof. Set $I = a_1 R \cup a_2 R \cup a_3 R \cup \cdots$. Then I is an ideal of R (Question 4, Exercises 14). Since R is a PID, $I = aR$ for some $a \in I$. Since $a \in I$ we have $a \in a_n R$ for some n. Thus by 14.8 $aR \subseteq a_n R$. However, $a_n R \subseteq I = aR$. Therefore $a_n R = aR$. Since $a \in a_n R$ we have $a \in a_s R$ for all $s \geq n$. It follows as above that for all such s we have $a_s R = aR = a_n R$. ∎

Note that by a 'non-trivial element' we mean an element which is not zero and which is not a unit.

Lemma 14.10. *Let R be a PID. Then every non-trivial element of R has an irreducible factor.*

Proof. Let x be a non-trivial element of R. Thus $x \neq 0$ and x is not a unit of R. Therefore $xR \neq 0$ and $xR \neq R$. If x is irreducible then x itself is the required factor and we can stop. Suppose that x is reducible (i.e. not irreducible). Since x is non-trivial, we have $x = a_1 b_1$ for some non-units $a_1, b_1 \in R$. Thus $x \in a_1 R$ so that $xR \subseteq a_1 R$. Suppose that $xR = a_1 R$. Then $a_1 \in xR$. Hence $a_1 = xr$ for some $r \in R$. Thus $a_1 = xr = a_1 b_1 r$ and so $b_1 r = 1$. This contradicts the fact that b_1 is not a unit of R. Therefore $xR \subsetneq a_1 R$. Since a_1 is not a unit of R, we can repeat the argument with a_1 in place of x. Suppose that a_1 is reducible. Then $a_1 = a_2 b_2$ for some non-units $a_2, b_2 \in R$. As before we have $a_1 R \subsetneq a_2 R$ and so on. By 14.9 this process must stop after a finite number of steps. When it does, we have an irreducible factor a_n of x. ∎

Lemma 14.11.   *Let R be a PID and let x be a non-trivial element of R. Then x is a product of irreducible elements of R.*

Proof.   By 14.10 there is an irreducible element $a_1 \in R$ such that $x = a_1 x_1$ for some $x_1 \in R$. If x_1 is a unit of R then $x = a_1 x_1$ with a_1 irreducible and x_1 is a unit, so that x is irreducible (see Question 1, Exercises 14) and we can stop. Suppose that x_1 is not a unit of R. Then there is an irreducible element $a_2 \in R$ such that $x_1 = a_2 x_2$ for some $x_2 \in R$. Suppose that x_2 is a unit of R. Then $x = a_1 \cdot a_2 x_2$ where a_1 and $a_2 x_2$ are irreducible elements of R. Now suppose that x_2 is not a unit of R. Then as in the proof of 14.10 we have $x_1 R \subsetneq x_2 R$ because $x_1 = a_2 x_2$ where a_2 is not a unit of R. We can continue in this way to obtain $x_1 R \subsetneq x_2 R \subsetneq x_3 R \subsetneq \cdots$. By 14.9 this process must stop after a finite number of steps. When it stops we have irreducible elements $a_1, a_2, \ldots, a_n \in R$ such that $x = a_1 a_2 \cdots a_n x_n$ for some unit $x_n \in R$. Since $a_n x_n$ is an irreducible element of R, this gives x as a product of irreducible elements of R.   ∎

Proposition 14.12.   *Let R be a PID and let x be a non-trivial element of R. Then xR is a maximal ideal of R if and only if x is an irreducible element of R.*

Proof.   Suppose that x is irreducible. Then x is not a unit and so $xR \neq R$. Let I be an ideal of R such that $I \supsetneq xR$. We have $I = aR$ for some $a \in R$. Hence $x \in aR$ so that $x = ar$ for some $r \in R$. Since x is irreducible, it follows that either a or r is a unit of R. However, $xR \neq aR$ and so x cannot be of the form au where u is a unit of R (Question 2, Exercises 14). Hence r is not a unit of R. Therefore a is a unit of R and so $aR = R$. Thus R is the only ideal which strictly contains xR. It follows that xR is a maximal ideal of R.

Conversely, suppose that xR is a maximal ideal of R and that $x = ar$ for some $a, r \in R$. By 14.8 we know that $aR \supseteq xR$. Since xR is a maximal ideal we must have either $aR = xR$ or $aR = R$. If $aR = xR$ then $a = xu$ for some unit $u \in R$, so that $a = aru$ and hence r is a unit of R with inverse u. If $aR = R$ then a is a unit of R. Thus x does not have a proper factorization, and so x is an irreducible element of R.   ∎

Lemma 14.13.   *Let R be a PID and let x be an irreducible element of R. Then x is a prime element of R.*

Proof.   Let a and b be elements of R such that $x \mid ab$. We must show that $x \mid a$ or $x \mid b$. Suppose that x does not divide a. Let $I = \{as + xt \mid s, t \in R\}$. It is easily seen that I is an ideal of R. Clearly $a \in I$ and $xR \subseteq I$. Since x does not divide a, by 14.8 we know that $a \notin xR$. Therefore $xR \subsetneq I$. Now by

14.12 xR is a maximal ideal of R. Hence $I = R$. So $1 \in I$ and we have $1 = as' + xt'$ for some $s', t' \in R$. Thus $b = abs' + xbt'$, where x divides both abs' and xbt'. Hence $x \mid b$.   ∎

Theorem 14.14. *Every PID is a UFD.*

Proof. Let R be a PID. By 14.11 every non-trivial element of R is a product of irreducible elements of R. The uniqueness of the factors in this product can be proved as in 6.12. By 14.13 the irreducible elements of R are prime.   ∎

Comments 14.15. Let R be a PID. Roughly speaking, by 14.12 we know that the maximal ideals of R are of the form aR where a is an irreducible element of R. Strictly speaking, we should also have assumed that R is not a field, because if R is a field then it is trivially a PID and $\{0\}$ is the only maximal ideal of R. The important examples of PIDs such as $\mathbb{Z}$ and $F[X]$ where F is a field are not themselves fields. By 14.12 the maximal ideals of $\mathbb{Z}$ are of the form $p\mathbb{Z}$ for some prime p. Let F be a field and set $R = F[X]$. We know by 14.12 that the maximal ideals of R are of the form aR, where a is an irreducible element of R. By 13.40 we know that R/aR is a field for such an element a. Thus we can construct new fields by starting with a known field F and forming factor rings of $F[X]$ by ideals determined by irreducible elements of $F[X]$.

Example 14.16. We shall construct a field with four elements. Set $F = \mathbb{Z}/2\mathbb{Z}$ and $R = F[X]$. We need an irreducible quadratic polynomial in R, and in fact there is only one such polynomial, namely $a = X^2 + X + 1$ (which is clearly irreducible since the equation $X^2 + X + 1 = 0$ has no solution in F). Set $I = aR$. We shall find the cosets of aR in R. Let $b \in R$. Then $b = aq + r$ for some $q, r \in R$ where either $r = 0$ or $\partial(r) < \partial(a)$. However, $\partial(a) = 2$. Hence $r = uX + v$ for some $u, v \in F$. There are four possibilities for r, namely $0, 1, X, 1 + X$. We have $b - r = aq$ so that $b - r \in I$. Hence by 13.24, we have $b + I = r + I$. Thus there are at most four cosets, namely $0 + I, 1 + I, X + I, 1 + X + I$. These cosets are distinct. In order to justify this we note that, because a is quadratic, the non-zero elements of aR have degree at least 2. Therefore the difference between any two of the elements $0, 1, X, 1 + X$ belongs to I if and only if it is 0. Thus R has precisely four distinct cosets with respect to the ideal aR and the ring R/I has four elements. Also since a is irreducible it follows (as stated in 14.15) that R/I is a field. We shall use the symbols $0, 1, x, 1 + x$ to denote the elements of R/I corresponding to the cosets $I, 1 + I, X + I$ and $1 + X + I$ respectively. Since R/I has order 2^2 we know that it must have characteristic 2. We can verify this by direct calculation: $(1 + I) + (1 + I) = 2 + I = 0 + I$, i.e. in R/I we have $1 + 1 = 0$. We have $(X + I)(X + I) = X^2 + I = 1 + X + I$, because

$$X^2 - (1 + X) = X^2 + X + 1 = a \in I.$$

Therefore in R/I we have $x^2 = 1 + x$. Thus the four elements $0, 1, x, 1 + x$ form a field in which the addition and multiplication are determined by the rules: $1 + 1 = x + x = (1 + x) + (1 + x) = 0$; $x^2 = 1 + x$. ∎

Theorem 14.17. *Let $R = F[X]$ where F is the field $\mathbb{Z}/p\mathbb{Z}$ for some prime p. Let a be an irreducible element of R with $\partial(a) = n$. Then R/aR is a field with p^n elements.*

Proof. Let $b \in R$. We have $b = aq + r$ for some $q, r \in R$ where either $r = 0$ or $\partial(r) < \partial(a)$. Then the cosets $b + aR$ and $r + aR$ are equal. Since $\partial(a) = n$ we have $r = f_0 + f_1 X + \cdots + f_{n-1} X^{n-1}$ for some $f_i \in F$. There are p possibilities for each f_i, so that there are p^n possibilities for r. Let these possible values of r be $r_1, r_2, \ldots, r_q$ where $q = p^n$. Then the cosets $r_i + aR$ are distinct and are the only cosets of aR in R. The distinctness of the cosets follows from the fact that the non-zero elements of aR have degree at least n, so that $r_i - r_j \in aR$ if and only if $r_i - r_j = 0$. Therefore there are q cosets of the maximal ideal aR in R. Hence R/aR is a field with q elements. ∎

Example 14.18. We shall construct a field with eight elements. Let $R = F[X]$ where $F = \mathbb{Z}/2\mathbb{Z}$. We want an irreducible cubic polynomial in R and so we can try taking $f(X) = X^3 + X + 1$. This is irreducible since the equation $f(X) = 0$ has no solution in F. Set $a = f(X)$. In the notation of 14.17 we have $p = 2$ and $n = 3$. Therefore R/aR is a field with 2^3 elements. As was shown in the proof of 14.17, the elements of R/aR can be represented as polynomials in x of degree less than 3 with coefficients in $\mathbb{Z}/2\mathbb{Z}$. The 8 elements of R/aR are $0, 1, x, 1 + x, x^2, x^2 + 1, x^2 + x, x^2 + x + 1$. We have $y + y = 0$ for all $y \in R/aR$. The multiplication in R/aR is governed by the rule $x^3 + x + 1 = 0$, i.e. $x^3 = -x - 1$, i.e. $x^3 = x + 1$. This is so because the action of forming R/aR has 'forced' us to have $f(X) = 0$, i.e. $X^3 + X + 1 = 0$. ∎

Comments 14.19. Let p be a prime number and n a positive integer. If we can find an irreducible polynomial of degree n over $\mathbb{Z}/p\mathbb{Z}$ then as shown in 14.17, we can construct a field with p^n elements. For small values of p and n we can usually find such a polynomial by trial and error. It is much more difficult to show that such a polynomial exists in general. We shall do this in Chapter 15.

We end this chapter by resolving a problem which was raised in 11.2(b). This is to show that given a field F and a polynomial $f(X)$ with coefficients in F, there is a field K which contains F and which also contains a root of the equation $f(X) = 0$.

Theorem 14.20 (Kronecker).   *Let $f(X)$ be a non-constant polynomial with coefficients in a field F. Then there exists a field K such that*

(i) *K contains F as a subfield, and*

(ii) *There is an element $t \in K$ such that $f(t) = 0$.*

Proof.   Set $R = F[X]$. We have $f(X) = f_1(X)f_2(X)\cdots f_k(X)$ for some irreducible elements $f_i(X) \in R$. We shall show that there is an extension K of F such that $f_1(t) = 0$ for some $t \in K$. Then $f(t) = 0$ also. Therefore without loss of generality we may suppose that $f(X)$ is irreducible over F.

Set $a = f(X)$ and $K = R/aR$. Then by 14.12 K is a field. Let $\lambda: R \to K$ be the canonical epimorphism, i.e. for each $r \in R$ we define $\lambda(r)$ to be the coset $r + aR$. We know that λ is a surjective homomorphism with ker $\lambda = aR$. Also, for $u_1, u_2 \in F$,

$$\lambda(u_1) = \lambda(u_2) \Rightarrow u_1 + aR = u_2 + aR \Rightarrow u_1 - u_2 \in aR \Rightarrow u_1 = u_2$$

since the non-zero elements of aR are polynomials of degree at least one. Thus the restriction of λ to F is injective and so λ gives an isomorphism between F and a subfield of K. We shall identify this subfield with F by writing $\lambda(u) = u$ for all $u \in F$. Then K can be considered an extension of F. Set $t = \lambda(X) = X + aR$. We have $a = f(X) = u_0 + u_1 X + \cdots + u_n X^n$ for some $u_i \in F$. Therefore

$$f(t) = u_0 + u_1 t + \cdots + u_n t^n = \lambda(u_0) + \lambda(u_1)\lambda(X) + \cdots + \lambda(u_n)(\lambda(X))^n$$
$$= \lambda(u_0 + u_1 X + \cdots + u_n X^n) = \lambda(f(X)) = \lambda(a) = 0$$

since ker $\lambda = aR$.   ∎

Exercises 14

1. Let a be an irreducible element of an integral domain R and let u be a unit of R. Show that au is an irreducible element of R.

2. Let R be an integral domain and $a, b \in R$. Show that $aR = bR$ if and only if $a = bu$ for some unit u of R.

3. Let F be a field and set $R = F[X, Y]$. Let I be the ideal of R which consists of those elements of R which have zero constant term. Show that the ideal I is not principal.

4. Let R be a ring with a sequence of ideals $I_1, I_2, \ldots$ such that

$$I_1 \subseteq I_2 \subseteq I_3 \subseteq \cdots .$$

Set $I = \bigcup_{i=1}^{\infty} I_i$. Prove that I is an ideal of R.

5. Let K be a field with four elements as in 14.16. Show that the equation $X^2 + X + 1 = 0$ has two distinct solutions in K.

6. Let K be a field with eight elements as in 14.18. Show that the equation $X^3 + X + 1 = 0$ has three distinct solutions in K.

7. Let p be a prime number. Prove that there are $p(p-1)/2$ monic quadratic polynomials which are irreducible over $\mathbb{Z}/p\mathbb{Z}$. [*Hint:* Count those which are reducible.] Hence show that there is a field with p^2 elements.

8. Set $K = R/aR$ where $R = (\mathbb{Z}/3\mathbb{Z})[X]$ and $a = X^2 + 1$. Prove that K is a field with nine elements. Verify that the multiplicative group of non-zero elements of K is cyclic.

9. Let K be an extension of a field F. Suppose that $t \in K$ is algebraic over F with minimum polynomial $m(X)$. Set $R = F[X]$ and $a = m(X)$. Show that $F(t) \cong R/aR$.

15 Finite fields

We shall show in this chapter that for each prime p and positive integer n there is one and only one field with p^n elements. This field is sometimes called the *Galois field* of order p^n and is denoted by $GF(p^n)$. We shall prove the existence of $GF(p^n)$ by showing that there is an irreducible polynomial of degree n over the field $\mathbb{Z}/p\mathbb{Z}$ (see 14.17). Finite fields have practical applications. The interested reader should consult Biggs (1985) for information in this area.

Notation 15.1. p will denote a prime number, n a positive integer. Let $q = p^n$ and let $g(n)$ denote the number of monic polynomials of degree n which are irreducible over the field $\mathbb{Z}/p\mathbb{Z}$.

The notation of 15.1 will be used from now on without further explanation. Our major aim is to show that $g(n) \neq 0$, i.e. that there is at least one monic irreducible polynomial of degree n over the field $\mathbb{Z}/p\mathbb{Z}$. Unfortunately, there does not seem to be a convenient way of expressing $g(n)$ directly in terms of n. However, we shall get the desired information from the number theoretic formula $p^n = \sum_{d \mid n} g(d)d$ where the sum is taken over all positive factors d of n (see 15.7). For example, when $n = 12$ this formula gives $p^n = g(1) + 2g(2) + 3g(3) + 4g(4) + 6g(6) + 12g(12)$.

Lemma 15.2. *Let d be a positive factor of n and let $f(X)$ be a monic polynomial of degree d which is irreducible over $\mathbb{Z}/p\mathbb{Z}$. Then $f(X) \mid X^q - X$.*

Proof. Set $R = (\mathbb{Z}/p\mathbb{Z})[X]$ and $a = f(X)$. Then R/aR is a field with p^d elements (14.17). Set $t = p^d$. For all $w \in R/aR$ we have $w^t - w = 0$ (Question 4, Exercises 9). Set $I = aR$ and let $r \in R$. In coset notation the condition $w^t - w = 0$ becomes $(r + I)^t - (r + I) = I$, i.e. $r^t - r + I = I$ and so $r^t - r \in I$. In particular, $X^t - X \in I = aR$. Thus $a \mid X^t - X$. We wish to show that $a \mid X^q - X$. Now $X^{t-1} - 1 \mid X^{q-1} - 1$ (see Question 1, Exercises 15). Hence $X^t - X \mid X^q - X$. Since $a \mid X^t - X$ we now have $a \mid X^q - X$. ∎

Lemma 15.3. *Let $f(X)$ be a monic irreducible polynomial of degree d over $\mathbb{Z}/p\mathbb{Z}$. Suppose that $f(X) \mid X^q - X$. Then $d \mid n$.*

Proof. Set $R = (\mathbb{Z}/p\mathbb{Z})[X]$, $a = f(X)$, $K = R/aR$ and $t = p^d$. Then by 14.17 K is a field with t elements. Let x denote the coset $X + aR$. By assumption $a \mid X^q - X$. Hence $X^q - X \in aR$. It follows by Proposition 13.24 that $X^q + aR = X + aR$. Hence $(X + aR)^q = X + aR$, i.e. $x^q = x$. By 2.1 we have $n = ds + r$ for some integers r and s where $0 \leq r < d$. Set $r' = p^r$. We have $p^n = (p^d)^s p^r$, i.e. $q = t^s r'$. Since K has t elements we have $x^t = x$ (Question 4, Exercises 9). Hence $(x^t)^t = x^t$ and so $(x^t)^t = x$. Continuing this way we find that $x^i = x$ if $i = t^j$ for some non-negative integer j. In particular $x^i = x$ when $i = t^s$. With this value of i we have $q = ir'$. As above we have $x = x^q$. Hence $x = x^q = (x^i)^{r'} = x^{r'}$.

Let $L = \{w \in K \mid w^{r'} = w\}$. It can be shown as in the proof of 10.11 that L is a subfield of K. This is true even when $r = 0$, i.e. $r' = 1$, because then $L = K$. However, $x \in L$. Therefore $L = K$ (see Question 2, Exercises 15). Thus K has t elements and they are all roots of the equation $Y^{r'} - Y = 0$. But $d > r$ and hence $p^d > p^r$ so that $t > r'$. Therefore by 5.11 $Y^{r'} - Y$ is the zero polynomial. Thus $r' = 1$ which gives $r = 0$. Therefore $d \mid n$. ∎

Definition 15.4. Let F be a field and

$$f(X) = a_n X^n + a_{n-1} X^{n-1} + \cdots + a_1 X + a_0$$

a polynomial with $a_i \in F$. Then

$$f'(X) = na_n X^{n-1} + (n-1)a_{n-1} X^{n-2} + \cdots + a_1$$

is called the *formal derivative* of $f(X)$.

Thus $f'(X)$ is the same as the derivative obtained by differentiation in calculus. The usual rules of calculus apply. Thus, for example, if $f(X)$ and $g(X)$ are two polynomials then $(f(X)g(X))' = f'(X)g(X) + f(X)g'(X)$.

Lemma 15.5. *Let $f(X)$ be a non-constant polynomial over $\mathbb{Z}/p\mathbb{Z}$. Then $(f(X))^2$ does not divide $X^q - X$.*

Proof. Suppose to the contrary that $X^q - X = (f(X))^2 h(X)$ for some polynomial $h(X)$ with coefficients in $\mathbb{Z}/p\mathbb{Z}$. Differentiating both sides of this equation gives

$$qX^{q-1} - 1 = 2f(X)f'(X)h(X) + (f(X))^2 h'(X).$$

Therefore $f(X) \mid qX^{q-1} - 1$. However, $qX^{q-1} = 0$ since $p \mid q$ and we are working over $\mathbb{Z}/p\mathbb{Z}$. Hence $f(X) \mid -1$, which is a contradiction since $f(X)$ is not a constant. ∎

Theorem 15.6. *Working over* $\mathbb{Z}/p\mathbb{Z}$ *we have* $X^q - X = f_1(X)f_2(X)\cdots f_k(X)$ *for some positive integer* k, *where*

(i) *each* $f_i(X)$ *is a monic irreducible polynomial;*

(ii) $f_i(X) \neq f_j(X)$ *if* $i \neq j$;

(iii) $\partial(f_i) \mid n$ *for all* i;

(iv) *if* $h(X)$ *is a monic irreducible polynomial such that* $\partial(h) \mid n$ *then* $h(X) = f_i(X)$ *for some* i.

Proof. Since $(\mathbb{Z}/p\mathbb{Z})[X]$ is a UFD (2.9 and 6.12) we can write $X^q - X$ uniquely in the form $f_1(X)f_2(X)\cdots f_k(X)$, where the $f_i(X)$ are monic irreducible polynomials. The theorem now follows easily from 15.2, 15.3 and 15.5.   ∎

Corollary 15.7. *In the notation of* 15.1 *we have* $q = \sum_{d \mid n} g(d)d$, *where* d *ranges over the positive factors of* n.

Proof. We write $X^q - X = f_1(X)f_2(X)\cdots f_k(X)$ as in 15.6. Then $q = \partial(X^q - X) = \partial(f_1) + \partial(f_2) + \cdots + \partial(f_k)$. For each i, $\partial(f_i) \mid n$, and for each positive factor d of n there are $g(d)$ polynomials $f_i(X)$ with $\partial(f_i) = d$. The result is now immediate.   ∎

Theorem 15.8. *Working over* $\mathbb{Z}/p\mathbb{Z}$, *there is at least one monic irreducible polynomial of degree* n.

Proof. In the notation of 15.1 we must show that $g(n) \neq 0$. The polynomial X is irreducible of degree 1, so that the result is true for $n = 1$. From now on we suppose that $n \geq 2$ and that $g(n) = 0$. We shall obtain a contradiction. As in 15.7 we have

$$p^n = \sum_{d \mid n} g(d)d, \text{ where the sum is taken over all positive factors } d \text{ of } n.$$

$$(*)$$

We have $g(n) = 0$ by assumption. Also $g(1) = p$ because the monic irreducible polynomials of degree 1 are of the form $X + a$ for $a \in \mathbb{Z}/p\mathbb{Z}$.

Let r be the largest integer such that $r \leq n/2$. Thus $r = n/2$ if n is even and $r = (n-1)/2$ if n is odd. Then r is at least 1 because n is at least 2. Let d be a positive factor of n with $d \neq 1$ and $d \neq n$. (If n is a prime number then there are no such factors d.) We have $dz = n$ for some positive integer z. Since $d \neq n$ we have $z \neq 1$ and so $z \geq 2$. Hence $d \leq n/2$. Therefore $d \leq r$ and we have $2 \leq d \leq r$. Thus there are at most $r - 1$ such factors d. Also $g(d) \leq p^d$ since the total number of monic polynomials of degree d over $\mathbb{Z}/p\mathbb{Z}$ is p^d. Hence $g(d)d \leq p^d d \leq p^r r$. Since there are at most $r - 1$ such factors d, we have $\sum g(d)d \leq (r-1)p^r r$ where the sum is taken over all positive factors d of n with $d \neq 1$ and $d \neq n$.

Since r is a positive integer we have $p \leq p^r$. Therefore it follows from $(*)$ that

$$p^n = g(1) + \sum_{\substack{d \mid n \\ d \neq 1 \\ d \neq n}} g(d)d + g(n)n$$

$$= p + \sum_{\substack{d \mid n \\ d \neq 1 \\ d \neq n}} g(d)d + 0 \leq p^r + (r-1)p^r r = (r^2 - r + 1)p^r.$$

Hence we have

$$p^{n-r} \leq r^2 - r + 1.$$

However, $r \leq n - r$. Hence $p^r \leq r^2 - r + 1$. Since p is a prime number we have $p \geq 2$. Therefore $2^r \leq r^2 - r + 1$. This is the desired contradiction, since in fact it is easy to show that $2^r > r^2 - r + 1$ for every positive integer r (see Question 3, Exercises 15). Thus $g(n) \neq 0$. ∎

Corollary 15.9. *There is a field with p^n elements.*

Proof. This readily follows from 15.8 and 14.17. ∎

Theorem 15.10. *Let K and L be fields with p^n elements. Then K is isomorphic to L.*

Proof. Set $F = \mathbb{Z}/p\mathbb{Z}$. Without loss of generality we may suppose that F is a subfield of both K and L (9.13). We have $K = F(t)$ for some element $t \in K$ which is algebraic of degree n over F (Question 11, Exercises 11). Let $m(X)$ be the minimum polynomial of t over F. Set $R = F[X]$ and $a = m(X)$. Then $K \cong R/aR$ (Question 9, Exercises 14).

We have $X^q - X = m(X)f(X)$ for some polynomial $f(X)$ with coefficients in F (15.2). Let $w \in L$. Then $w^q = w$ (Question 4, Exercises 9). Hence $m(w)f(w) = w^q - w = 0$. Thus $m(w) = 0$ or $f(w) = 0$. The q possible values of w cannot all satisfy $f(w) = 0$ because $\partial(f) < q$. Therefore there is an element $u \in L$ such that $m(u) = 0$. Since $m(X)$ is irreducible and $m(u) = 0$, $m(X)$ must be the minimum polynomial of u over F. Thus

$F(u)$ is a subfield of L with $[F(u):F] = n$ (11.10). As F has p elements we know that $F(u)$ has p^n elements. Therefore $F(u) = L$. Hence $L \cong R/aR$ (Question 9, Exercises 14). Therefore $K \cong L$ since they are both isomorphic to R/aR.   ■

Summary 15.11.   We have shown in 15.9 and 15.10 that for each prime number p and positive integer n there is, up to isomorphism, exactly one field with p^n elements. This field has exactly one subfield with p^d elements for each positive factor d of n (10.11). The automorphism group of the field is cyclic of order n and is generated by the Frobenius automorphism (Question 11, Exercises 13).

Exercises 15

1. Let n, r, d be positive integers such that $n = rd$. Set $q = p^n$ and $t = p^d$. Prove that $X^{t-1} - 1 \mid X^{q-1} - 1$.

2. In the notation of the proof of 15.3 prove that $K = L$.
   (*Hint:* As in 14.16 and in the proof of 14.17, the elements of K can be represented as polynomial expressions in x with coefficients in $\mathbb{Z}/p\mathbb{Z}$.)

3. Prove that $2^r > r^2 - r + 1$ for every positive integer r.

Solutions to selected exercises

Solutions 1

1. $n \mid a - b$ and $n \mid c - d$. So $n \mid (a + c) - (b + d)$. We also have $ac - bd = a(c - d) + (a - b)d$. Therefore $n \mid ac - bd$.

2. If $1 = 0$ then for all $x \in R$ we have $x = 1x = 0x = 0$.

8. We have $(w - 1)(w^2 + w + 1) = w^3 - 1 = 0$. Since $w - 1 \neq 0$ we obtain $w^2 + w + 1 = 0$. The only ring axiom which is not obvious is closure under multiplication. Let $a, b, c, d \in \mathbb{Z}$. We have

$$(a + bw)(c + dw) = ac + (ad + bc)w + bdw^2.$$

Now $w^2 = -w - 1$ so that

$$(a + bw)(c + dw) = (ac - bd) + (ad + bc - bd)w,$$

which is of the right form. We have

$$(x + y)(x + wy)(x + w^2 y)$$
$$= x^3 + (1 + w + w^2)x^2 y + (w^3 + w^2 + w)xy^2 + w^3 y^3 = x^3 + y^3$$

since $w^3 = 1$ and $w^2 + w + 1 = 0$.

9. One can simply test whether any of the elements 0, 1, 2, 3, 4, 5, of $\mathbb{Z}/6\mathbb{Z}$ satisfies the given equation. The solutions are (i) 2 and 5, (ii) no solutions, (iii) 1, 2, 4 and 5. Note that there are four solutions to this quadratic equation.

   The usual formula for quadratics involves dividing by 2 (and taking square root). So it requires a multiplicative inverse for 2. However, there is no w in $\mathbb{Z}/6\mathbb{Z}$ with $2w = 1$.

Solutions 2

2. Suppose firstly that R is an integral domain. Let $f(X)$ and $g(X)$ be non-zero polynomials in $R[X]$. Set $n = \partial(f)$ and $k = \partial(g)$. Let a_n, b_k be the leading coefficients of $f(X)$ and $g(X)$ respectively. We have $a_n \neq 0$ and $b_k \neq 0$. Since R is an integral domain we have $a_n b_k \neq 0$.

Now $a_n b_k$ is the coefficient of X^{n+k} in $f(X)g(X)$. Therefore $f(X)g(X) \neq 0$. Thus $R[X]$ is an integral domain.

Conversely, suppose that $R[X]$ is an integral domain. Let a, b be non-zero elements of R. Then a, b are non-zero in the integral domain $R[X]$. Hence $ab \neq 0$ and R is an integral domain.

3. This is similar to the first part of solution 2.

4. Let R be a ring and let U be the set of units of R. Clearly $1 \in U$. Let $u, v \in U$. Then u and v have inverses u^{-1} and v^{-1} in R and $u^{-1}v^{-1}$ is the inverse for uv. Therefore $uv \in U$. Also u^{-1} is a unit of R, the inverse being u. Finally the multiplication in U is associative because the multiplication in R is associative.

7. Take $a(X) = X$ and $b(X) = 2$. If the division algorithm were valid we would have polynomials $q(X)$ and $r(X)$ with integer coefficients such that $a(X) = b(X)q(X) + r(X)$ with either $r(X) = 0$ or $\partial(r) < \partial(2)$. This would force $r(X) = 0$ and hence $a(X) = b(X)q(X)$. Thus $X = 2q(X)$, which is a contradiction since the coefficients of $q(X)$ are integers.

9. If $d = 2$ or 3 or -2 then for all non-zero $a, b \in R$ we have $N(ab) = N(a)N(b) \geq N(a)$ since $N(b)$ is a positive integer (in these cases $N(x) = 0$ if and only if $x = 0$).
   (i) $d = 2$. Let $x, y \in R$ with $y \neq 0$. Then we have $x = g + h\sqrt{2}$ and $y = s + t\sqrt{2}$ for some g, h, s and $t \in \mathbb{Z}$. Hence

   $$x/y = (g + h\sqrt{2})/(s + t\sqrt{2})$$
   $$= (g + h\sqrt{2})(s - t\sqrt{2})/(s^2 - 2t^2) = u + v\sqrt{2}$$

   for some rational numbers u and v. Let a and b be the nearest integers to u and v respectively. Set $q = a + b\sqrt{2}$ and $r = x - qy$. We want $N(r) < N(y)$, i.e. $N(r/y) < 1$, i.e. $N(x/y - q) < 1$. We have $x/y - q = (u - a) + (v - b)\sqrt{2}$, so

   $$N(x/y - q) = |(u - a)^2 - 2(v - b)^2|.$$

   Now

   $$(u - a)^2 - 2(v - b)^2 \leq (u - a)^2 \leq 1/4;$$

   and

   $$(u - a)^2 - 2(v - b)^2 \geq -2(v - b)^2 \geq -2/4 = -1/2.$$

   Therefore $N(x/y - q) < 1$ as required.

(ii) $d = 3$. This is very similar to $d = 2$ except that at the end we get $-3/4$ rather than $-2/4$.

(iii) $d = -2$. This is similar but slightly easier because we can delete the absolute value signs in the definition of N and have

$$N(x/y - q) = (u - a)^2 + 2(v - b)^2 \le 1/4 + 2/4 = 3/4.$$

10. $w^2 = (1 + 2\sqrt{(-3)} - 3)/4 = 1/2(\sqrt{(-3)} - 1) = w - 1$. To show that A is a subring of $\mathbb{C}$, the only thing which is not obvious is closure under multiplication. Let $a, b, c, d \in \mathbb{Z}$. Then

$$(a + bw)(c + dw) = ac + (ad + bc)w + bdw^2$$

$$= ac - bd + (ad + bc + bd)w.$$

Hence A is closed under multiplication. For $r \in A$ we have $r = a + bw$ for some $a, b \in \mathbb{Z}$. Clearly $N(r)$ is never negative because $N(r) = |r|^2$. Also

$$N(r) = N(a + bw) = N((a + b/2 + ib\sqrt{3}/2)$$

$$= (a + b/2)^2 + 3b^2/4 = a^2 + ab + b^2,$$

which is an integer. Let z be an arbitrary complex number with $z = x + iy$ where $x, y \in \mathbb{R}$. We want $z = u + vw$ for some real numbers u and v, i.e. $x + iy = u + v/2 + iv\sqrt{3}/2$. So we can take $v = 2y/\sqrt{3}$ and $u = x - y/\sqrt{3}$. Changing notation, let $x, y \in A$ with $y \ne 0$. Then $x/y = u + vw$ for some $u, v \in \mathbb{R}$. Let a and b be the nearest integers to u and v. Set $q = a + bw$ and $r = x - yq$. Then

$$N(r/y) = N(x/y - q) = N((u - a) + (v - b)w)$$

$$= (u - a)^2 + (u - a)(v - b) + (v - b)^2 \le 1/4 + 1/4 + 1/4 < 1.$$

Thus A satisfies the division algorithm with respect to N. Let r, s be non-zero elements of A. Then $N(rs) = N(r)N(s) \ge N(r)$ because $N(s)$ is a positive integer. Therefore A is Euclidean with respect to N. We have $N(1) = 1$. Hence to find the units of A we need to find all integers a and b such that $N(a + bw) = 1$, i.e. $a^2 + ab + b^2 = 1$. If ab is not negative then $1 = a^2 + ab + b^2$ if and only if either $a^2 = 0$ and $b^2 = 1$, or $a^2 = 1$ and $b^2 = 0$. Also $a^2 + ab + b^2 = (a + b)^2 - ab$. Hence if ab is negative and $(a + b)^2 - ab = 1$ we must have $ab = -1$ and $a + b = 0$, i.e. either $a = 1$ and $b = -1$, or $a = -1$ and $b = 1$. These solutions for a, b correspond to the six units given in the question.

Solutions 3

6. Let $r = (p - 1)/2$. Since p is odd, r is an integer. Working in $\mathbb{Z}/p\mathbb{Z}$ we have $a^{2r} = 1 = b^{2r}$ by Fermat's little theorem since, by assumption, neither a nor b is zero in $\mathbb{Z}/p\mathbb{Z}$. So $1 = (a^2)^r = (-b^2)^r = (-1)^r b^{2r} = (-1)^r$. Thus $(-1)^r = 1$ in $\mathbb{Z}/p\mathbb{Z}$. Thus $p \mid (1 - (-1)^r)$ in $\mathbb{Z}$. Since p does not divide 2, we must have $1 - (-1)^r = 0$. Hence r is even. Thus $r = 2n$ for some integer n. Therefore $(p - 1)/2 = 2n$ and $p = 4n + 1$.

7. Write $28 = 4 \cdot 7$ a product of relatively prime terms. Then $17^{500} \equiv 1^{500} \equiv 1 \pmod 4$ and $17^{500} \equiv (17^6)^{83} \cdot 17^2 \equiv 17^2 \equiv 3^2 \equiv 2 \pmod 7$. Thus we require x between 0 and 27 such that $x \equiv 1 \pmod 4$ and $x \equiv 2 \pmod 7$. Therefore $x = 9$.

8. As $4k - 1$ is odd, its prime factors are all odd. Suppose that all the prime factors of $4k - 1$ are congruent to 1 (mod 4). A product of numbers congruent to 1 is itself congruent to 1. Hence $4k - 1 \equiv 1 \pmod 4$. Thus $-1 \equiv 1 \pmod 4$ which is a contradiction. Therefore $4k - 1$ has an odd prime factor p which is not congruent to 1 (mod 4). Thus $p \equiv 3 \pmod 4$. Since p divides $4k - 1$, it is not possible that p divides k. Therefore if S is any finite set of primes congruent to 3 (mod 4) then there is at least one such prime not in S. Hence the set of all such primes is infinite.

Solutions 4

4. We prove this by induction on t. The case $t = 0$ is covered by Question 3. So let $t > 0$ and suppose that $n = a^2 + b^2 + c^2$. Since $4 \mid n$, in $\mathbb{Z}/4\mathbb{Z}$ we have $a^2 + b^2 + c^2 = 0$. However, each of a^2, b^2, c^2 is 0 or 1 in $\mathbb{Z}/4\mathbb{Z}$. As they add up to 0 and there are only three of them, they are all zero. Hence in $\mathbb{Z}$ the integers a, b, c are all even. Thus $a = 2u$, $b = 2v$, $c = 2w$ for some $u, v, w \in \mathbb{Z}$. Since

$$a^2 + b^2 + c^2 = 4^t(8k + 7),$$

we have

$$u^2 + v^2 + w^2 = 4^{t-1}(8k + 7).$$

Now by induction we may suppose that $4^{t-1}(8k + 7)$ cannot be the sum of three squares. Hence $4^t(8k + 7)$ is not the sum of three squares.

8. Let n be a positive integer. We can write $n = mk^2$ where m and k are positive integers and m has no repeated prime factors. Suppose that $m = a^2 + b^2 + c^2 + d^2$ for some $a, b, c, d \in \mathbb{Z}$. Then

$$n = (ak)^2 + (bk)^2 + (ck)^2 + (dk)^2.$$

Solutions 5

4. Let a be a non-zero element of R and $S = \{ab \mid b \in R\}$. Now R is an integral domain and $a \neq 0$ and so $ab = ac$ for $b, c \in R$ implies that $b = c$. Since R is finite it follows that S has as many elements as R. Hence $S = R$. Therefore $1 \in S$. Thus there exists $a' \in R$ such that $aa' = 1$ and R is a field.

6. The case in which $p = 2$ is easy, and so from now on we shall assume that p is odd. By 3.16 each $a \in F$ satisfies $f(a) = 0$. Thus $f(X) = 0$ has $p - 1$ distinct roots in F. However, multiplying out the brackets we see that $f(X)$ is actually a polynomial of degree at most $p - 2$. Hence by 5.11 the polynomial $f(X)$ must be identically zero, i.e. all its coefficients must be zero. Putting the constant term equal to zero we have $-1 - (p - 1)! = 0$ in F. Thus $(p - 1)! \equiv -1 \pmod{p}$.

Solutions 6

2. Let p be a prime element of an integral domain R and suppose that $p = ab$ for some $a, b \in R$. Certainly $p \mid ab$, so that $p \mid a$ or $p \mid b$. Without loss of generality suppose that $p \mid a$. Then $a = pr$ for some $r \in R$. Hence $p = ab = prb$ so that $1 = rb$. Thus b is a unit.

3. Let x be an irreducible element of a UFD R and suppose that $x \mid ab$ for some $a, b \in R$. We have $a = p_1 p_2 \cdots p_r$ and $b = q_1 q_2 \cdots q_s$, where each p_i and q_j is irreducible. Thus $p_1 p_2 \cdots p_r q_1 q_2 \cdots q_s$ is the unique factorization of ab into irreducible factors. However, $ab = xy$ for some $y \in R$. We have $y = w_1 w_2 \cdots w_t$ for some irreducible elements w_i. Thus $p_1 p_2 \cdots p_r q_1 q_2 \cdots q_s$ and $x w_1 w_2 \cdots w_t$ are both factorizations of ab into irreducible factors. Since R is a UFD, we know that $x = u p_i$ or $u q_j$ for some unit u of R. Hence x is a factor of a or b.

4.   (i) *Reflexive:* For all $a \in R$ we have $a = 1a$, so that a is an associate of a.

      Symmetric: Suppose that a is an associate of b. Then $a = ub$ for some unit u of R. Hence $b = u^{-1}a$. Now u^{-1} is a unit of R and so b is an associate of a.

      Transitive: Suppose that $a = ub$ and $b = vc$ where u, v are units of R. Then $a = uvc$ and uv is a unit of R.

  (ii) Suppose that a is an associate of b so that $a = ub$ for some unit $u \in R$. The equation $a = ub$ shows that $b \mid a$, and $b = u^{-1}a$ shows that $a \mid b$.

      Conversely suppose that $a \mid b$ and $b \mid a$. Thus $a = rb$ and $b = sa$ for some $r, s \in R$. If one of a or b is 0 then so is the other, and they are associates. Suppose that $a \neq 0$. Now $a = rb = rsa$ and so $rs = 1$. Thus $a = rb$ where r is a unit of R.

(iii) Suppose that $a \mid b$ and that c is an associate of a. Then $b = ra$ and $c = ua$ for some $r, u \in R$ with u a unit. Thus $b = ru^{-1}c$ with $ru^{-1} \in R$. Thus $c \mid b$.

(iv) Let g and h be HCFs of a and b. Then g divides both a and b so that $g \mid h$. By symmetry $h \mid g$. Therefore g and h are associates.

Let d be an associate of h. Since $h \mid a$ and b, so also does d. Also since $h \mid d$, any common factor of a and b divides h and hence divides d. Therefore d is an HCF of a and b.

6. Suppose that $h(X)$ is a common factor of 2 and X. Then $2 = f(X)h(X)$ and $X = g(X)h(X)$ for some polynomials $f(X), g(X)$ with integer coefficients. It follows from $2 = f(X)h(X)$ that both $f(X)$ and $h(X)$ are constants, say $f(X) = a$ and $h(X) = b$ for some $a, b \in \mathbb{Z}$. Thus $2 = ab$ so that b is one of the numbers $2, -2, 1, -1$. Now $X = g(X)b$ and since $g(X)$ has integer coefficients, we must have $b = 1$ or -1. Thus the only common factors of 2 and X are units.

Suppose that $1 = 2s(X) + Xt(X)$ for some polynomials $s(X)$ and $t(X)$ with integer coefficients. Equating the constant terms we find that $1 = 2s_0$ where s_0 is the constant term of $s(X)$. This is a contradiction because s_0 is an integer.

Solutions 7

4. We have $x_i = u_i/v_i$ for some $u_i, v_i \in R$ with $v_i \neq 0$. Set $b = v_1 v_2 \cdots v_n$ and $a_i = v_1 v_2 \cdots v_{i-1} u_i v_{i+1} \cdots v_n$. Then $x_i = a_i/b$ as required.

Solutions 8

2. (iv) $X^2 + 3 = (X - 2)(X + 2)$ over $\mathbb{Z}/7\mathbb{Z}$.

   (vi) $X^3 + 2 = (X - 1)^3$ over $\mathbb{Z}/3\mathbb{Z}$.

   (viii) Set $f(X) = X^5 - 10X + 1$. Put $X = Y - 1$. Check that $f(Y - 1)$ is irreducible by Eisenstein's criterion with $p = 5$.

   (xi) Over $\mathbb{Z}/2\mathbb{Z}$ we have $X^4 + 3X^3 - 3X - 2 = X(X^3 + X^2 + 1)$. The equation $X^3 + X^2 + 1 = 0$ has no solution over $\mathbb{Z}/2\mathbb{Z}$ so that $X^3 + X^2 + 1$ is irreducible.

3. The polynomial factorizes properly if and only if there is an integer x such that $x^2 \equiv 3 \pmod{79}$. In $\mathbb{Z}/79\mathbb{Z}$ we have $3^{39} = (3^4)^9 \cdot 3^3 = 81^9 \cdot 27 = 2^9 \cdot 27 = 512 \cdot 27 = 38 \cdot 27 = 1026 = (13 \cdot 79) - 1 = -1$.

Solutions 9

3. Let 1_K and 1_F denote the identity elements of K and F respectively. The non-zero elements of F form a subgroup of the group of non-zero

elements of K under multiplication. Hence the corresponding identity elements 1_K and 1_F are equal.

4. Let $x \in F^*$ and let k be the order of x as an element of F^*. We have $x^k = 1$. Since F^* has order $q - 1$ we know that $k \mid (q - 1)$. Thus $q - 1 = kr$ for some $r \in \mathbb{Z}$. Hence $x^{q-1} = (x^k)^r = 1^r = 1$. Therefore $x^q = x$, and this is also true for $x = 0$.

7. Binomial coefficients are positive integers so that $p!/r!(p-r)! = k$ for some $k \in \mathbb{Z}$. Thus $p! = r!(p-r)!k$. But $p \mid p!$. Hence p divides $r!$ or $(p-r)!$ or k. However, the integers $1, 2, \ldots, r$ are smaller than p and so are not divisible by p. Therefore p does not divide $r!$. Similarly p does not divide $(p-r)!$. Therefore $p \mid k$.

11. Suppose that $K = F$. Each element $k \in K$ can be written uniquely in the form $k = f \cdot 1$ with $f \in F$ (namely $f = k$). Thus the single element 1 forms a basis for K over F so that $[K : F] = 1$. Conversely suppose that $[K : F] = 1$. The non-zero element 1 of K forms a linearly independent set over F and hence forms a basis. Let $k \in K$. Then $k = f \cdot 1$ for some $f \in F$. Thus $k = f$ and $k \in F$. Hence $K = F$.

Solutions 10

2. (i) The complex numbers of the form $e^{2ki\pi/n}$, where k and n are integers with n positive; (ii) 1 and -1; (iii) 1 and -1.

5. After each value of n we shall give (a) all the elements of the group of units of $\mathbb{Z}/n\mathbb{Z}$ and (b) all those elements which generate the group as a cyclic group.
   $n = 2$: (a) 1 (b) 1;
   $n = 3$: (a) 1, 2 (b) 2;
   $n = 4$: (a) 1, 3 (b) 3;
   $n = 5$: (a) 1, 2, 3, 4 (b) 2, 3;
   $n = 6$: (a) 1, 5 (b) 5;
   $n = 7$: (a) 1, 2, 3, 4, 5, 6 (b) 3, 5;
   $n = 9$: (a) 1, 2, 4, 5, 7, 8 (b) 2, 5;
   $n = 10$: (a) 1, 3, 7, 9 (b) 3, 7;
   $n = 11$: (a) 1, 2, 3, \ldots, 10 (b) 2, 6, 7, 8;
   $n = 13$: (a) 1, 2, 3, \ldots, 12 (b) 2, 6, 7, 11;
   $n = 14$: (a) 1, 3, 5, 9, 11, 13 (b) 3, 5;
   $n = 17$: (a) 1, 2, 3, \ldots, 16 (b) 3, 5, 6, 7, 10, 11, 12, 14;
   $n = 18$: (a) 1, 5, 7, 11, 13, 17 (b) 5, 11;
   $n = 19$: (a) 1, 2, 3, \ldots, 18 (b) 2, 3, 10, 13, 14, 15.

7. Set $k = ab/(a, b)$. Note that k is a positive integer which is divisible by both a and b, and that $k < ab$. Let $x = (u, v)$ with $u \in \mathbb{Z}/a\mathbb{Z}, v \in \mathbb{Z}/b\mathbb{Z}$.

Since $o(u) \mid o(\mathbb{Z}/a\mathbb{Z}) = a$ we have $au = 0$ (this is the additive version of $u^a = 1$). Similarly $bv = 0$. Since $a \mid k$ and $b \mid k$ we have $ku = 0 = kv$ so that $kx = 0$. Therefore $o(x) \mid k$ so that $o(x) < ab$. Hence x does not generate G.

10. The integers from 1 to p^n which are not relatively prime to p^n are those which are divisible by p, namely $p, 2p, 3p, \ldots, p^{n-1}p$, and there are p^{n-1} of them. Hence $\varphi(p^n) = p^n - p^{n-1} = p^{n-1}(p-1)$.

11. If $n = p_1^{a_1} p_2^{a_2} \cdots p_r^{a_r}$, where the p_i are distinct primes and the a_i are positive integers, then

$$\varphi(n) = \varphi(p_1^{a_1})\varphi(p_2^{a_2}) \cdots \varphi(p_r^{a_r}) \text{ by 10.21}$$

$$= p_1^{a_1}(1 - 1/p_1)p_2^{a_2}(1 - 1/p_2) \cdots p_r^{a_r}(1 - 1/p_r) \text{ by Question 10}$$

$$= n(1 - 1/p_1)(1 - 1/p_2) \cdots (1 - 1/p_r).$$

Solutions 11

1. Suppose that $t \in F$. Then t is a root of the linear equation $X - t = 0$ which has coefficients in F. Therefore t is algebraic of degree 1 over F.

   Conversely suppose that t is algebraic of degree 1 over F. Let $m(X)$ be the minimum polynomial of t. Then $\partial(m) = 1$ and $m(X) = X + a$ for some $a \in F$. However $m(t) = 0$ so that $t = -a$ and $t \in F$.

4. Fix an element $t \in K$ with $t \notin F$. The three elements $1, t, t^2 \in K$ are linearly dependent over F. Therefore there are $a, b, c \in F$, not all zero, such that $a + bt + ct^2 = 0$. Suppose that $c = 0$. Then $bt = -a$. If $b \neq 0$ then $t = -a/b$, which is a contradiction since $t \notin F$. Hence $b = 0$. Therefore $a = -bt = 0$, which is a contradiction since not all a, b, c are zero. Hence $c \neq 0$. The standard formula for solving quadratics gives

$$t = \left(-b + (b^2 - 4ac)^{1/2}\right)/2c,$$

where $(b^2 - 4ac)^{1/2}$ is one of the square roots of $b^2 - 4ac$. Set $d = b^2 - 4ac$. Then $d \in F$ and $t = (-b + \sqrt{d})/2c$. Hence $t \in F(\sqrt{d})$ so that $F(t) \subseteq F(\sqrt{d})$. Also $\sqrt{d} = 2ct + b$ so that $\sqrt{d} \in F(t)$. Therefore $F(\sqrt{d}) \subseteq F(t)$ so that $F(\sqrt{d}) = F(t)$. As $[K:F] = 2$ and $t \notin F$, we have $K = F(t) = F(\sqrt{d})$.

6. We have $3 = (t - \sqrt{2})^2 = t^2 - 2t \cdot \sqrt{2} + 2$ and so $2t \cdot \sqrt{2} = t^2 - 1$. Hence $8t^2 = t^4 - 2t^2 + 1$. Set $m(X) = X^4 - 10X^2 + 1$. Then $m(t) = 0$. Suppose that $a \in \mathbb{Q}$ and $m(a) = 0$. Then a is an integer (why?) which

divides the constant term 1 of $m(X)$. Hence $a = 1$ or $a = -1$. However, $m(1) \neq 0 \neq m(-1)$ and so $m(X)$ has no linear factors over $\mathbb{Q}$. Suppose that $m(X)$ is the product of two quadratic factors over $\mathbb{Q}$. Without loss of generality we may suppose that

$$m(X) = (X^2 + aX + b)(X^2 + cX + d)$$

with $a, b, c, d \in \mathbb{Z}$. Equating coefficients gives $a + c = 0$, $b + d + ac = -10$, $ad + bc = 0$, $bd = 1$. Hence either $b = d = 1$ or $b = d = -1$. Also $c = -a$. Thus $-10 = b + d + ac = 2b - a^2$ and so $a^2 = 10 + 2b$. Hence $a^2 = 8$ or 12, which is a contradiction. Therefore $m(X)$ is irreducible over $\mathbb{Q}$ and is the minimum polynomial for t over $\mathbb{Q}$. Also $t^4 = 10t^2 - 1$ so that $t^5 = (10t^2 - 1)t = 10t^3 - t$.

7. Set $f(X) = X^n + f_{n-1}X^{n-1} + \cdots + f_1 X + f_0$. Equating constant terms and coefficients of X^{n-1} in the two representations of $f(X)$ gives $a_1 a_2 \cdots a_n = (-1)^n f_0$ and $a_1 + a_2 + \cdots + a_n = -f_{n-1}$.

8. (i) Since $f(a) = 0$ and $f(X)$ is irreducible over F, we know that $f(X)$ is the minimum polynomial for a over F. Therefore $[F(a):F] = \partial(f) = 3$.

   (ii) The equation $f(X) = 0$ has a solution in $F(a)$, namely a. Hence $f(X) = (X - a)g(X)$ for some polynomial $g(X)$ with coefficients in $F(a)$. Since $f(X) = (X - a)(X - b)(X - c)$ we have $g(X) = (X - b)(X - c)$ so that $g(b) = 0$. Therefore b is a root of the quadratic equation $g(X) = 0$ which has coefficients in $F(a)$.

   (iii) Let u be the coefficient of X^2 in $f(X)$. Then $u \in F$ and $a + b + c = -u$ (see solution 7 above). Hence $c = -(u + a + b)$ so that $c \in F(a, b)$. Therefore $F(a, b, c) = F(a, b)$.

   (iv) We have $[F(a, b, c):F] = [F(a, b):F(a)][F(a):F]$. We know that $[F(a):F] = 3$. Also b satisfies a quadratic equation over $F(a)$ so that the degree of the minimum polynomial for b over $F(a)$ is at most 2. Hence $[F(a, b):F(a)] = 1$ or 2.

11. Set $q = p^n$. Then $S(t)$ contains all the elements $0, 1, t, t^2, \ldots, t^{q-1}$. Hence $S(t) = F$ because $1, t, t^2, \ldots, t^{q-1}$ are the non-zero elements of F. We have $|F| = p^n$ and $|S| = p$. Hence $[F:S] = n$ (see the proof of 9.23). Therefore $[S(t):S] = n$ so that t is algebraic of degree n over S by 11.10.

Solutions 12

3. Let G be the multiplicative group of non-zero elements of $\mathbb{Z}/17\mathbb{Z}$. Then $|G| = 16$. So the order of 3 as an element of G is 1, 2, 4, 8 or 16.

However (mod 17) we have $3^2 \equiv 9$, $3^4 \equiv 81 \equiv 13 \equiv -4$ and $3^8 \equiv 16 \equiv -1$. Thus 3 does not have order 1, 2, 4 or 8. Therefore 3 has order 16 and so it generates G.

4. Let the circle centred at B passing through A cut AB at A and C. Let P be one of the points where the circle centred at A passing through C cuts the circle centred at C passing through A. Then PB is perpendicular to AB.

5. Suppose that $k = rs$ where r and s are positive integers with r odd. Set $u = 2^s$. Then $p = u^r + 1 = (u + 1)(u^{r-1} - u^{r-2} + \cdots - u + 1)$. However, p is prime and $u + 1$ is a positive integer with $u + 1 \neq 1$. Therefore $u + 1 = p$. Hence $u^r + 1 = u + 1$. Since $u \geq 2$ it follows that $r = 1$. Thus k has no proper odd factors and so k must be a power of 2.

6. A regular 4-gon, i.e. a square, can be inscribed in the unit circle by joining the points $(1, 0)$, $(0, 1)$, $(-1, 0)$, and $(0, -1)$. Each side of the square subtends an angle of $90°$ at the centre. These angles can be bisected and the points where the bisecting lines cut the unit circle, together with the vertices of the square, form the vertices of the regular 8-gon. By repeatedly bisecting angles in this way it is possible to construct the regular 2^k-gon for every positive integer $k \neq 1$.

7. Suppose that $q_1 w + q_2 w^2 + \cdots + q_{16} w^{16} = 0$ for some $q_i \in \mathbb{Q}$. Cancelling w gives $q_1 + q_2 w + \cdots + q_{16} w^{15} = 0$. However, $1, w, \ldots, w^{15}$ are linearly independent over $\mathbb{Q}$. Therefore $q_i = 0$ for all i. Thus the 16 elements $w, w^2, \ldots, w^{16}$ are linearly independent over $\mathbb{Q}$. Since $\mathbb{Q}(w)$ has dimension 16 as a vector space over $\mathbb{Q}$ it follows that $w, w^2, \ldots, w^{16}$ form a basis for this vector space.

Solutions 13

3. Let G be the set of all automorphisms of R with multiplication in G given by composition of functions. Taking $R = S = T$ in Question 2 shows that the product of two automorphisms of R is an automorphism of R. Hence G is closed under multiplication. The multiplication in G is associative because it is composition of functions. The identity function on R is the identity of G. Let g be an automorphism of R. Then the inverse function g^{-1} is also an automorphism of R, by Question 1. Hence g^{-1} is the inverse of g in G.

6. Let I be an ideal of the field F and suppose that I contains a non-zero element x. We have $xy = 1$ for some $y \in F$. However, $xy \in I$ since $x \in I$ and $y \in F$. Therefore $1 \in I$. Hence for any $z \in F$ we have $z = 1 \cdot z \in I$ and so $I = F$.

7. Suppose that p is a prime number. Then by 5.7 the ring $\mathbb{Z}/p\mathbb{Z}$ is a field and so by 13.40 $p\mathbb{Z}$ is a maximal ideal of $\mathbb{Z}$.

   Conversely, suppose that M is a maximal ideal of $\mathbb{Z}$. Then $M = n\mathbb{Z}$ for some $n \in \mathbb{Z}$. Clearly $n \neq 0$ since $n = 0$ gives $M = \{0\} \subsetneq 2\mathbb{Z} \subsetneq \mathbb{Z}$ contradicting the maximality of M. Since $n\mathbb{Z} = (-n)\mathbb{Z}$ we can also assume that n is positive. Also $n \neq 1$. However $\mathbb{Z}/n\mathbb{Z}$ is a field and so by 5.7 the integer n is prime.

8. Set $S = \mathbb{Z}/2\mathbb{Z}$ and define $f: R \to S$ as follows. Let $r \in R$. Set $f(r) = 0$ or 1 according as the constant term of r is even or odd. Then f is a surjective homomorphism with $\ker f = K$. Therefore by 13.42 $R/K \cong S$.

Solutions 14

1. Let $u \in R$ be a unit. Set $b = ua$. We must show that b is irreducible. We have $b \neq 0$ since $a \neq 0$. Also $a = u^{-1}b$ where a is not a unit. Hence b is not a unit. Suppose that $b = cd$ where c and d are non-units of R. Then $a = (u^{-1}c)d$ where $u^{-1}c$ and d are non-units of R. This is a contradiction since a is irreducible. Hence b is irreducible.

4. Since I_i contains the zero element of R so also does I. Let $a, b \in I$. Then there are positive integers j and k such that $a \in I_j$ and $b \in I_k$. Without loss of generality we may suppose that $j \leq k$. Since $I_j \subseteq I_k$ it follows that $a \in I_k$ and $b \in I_k$. Hence $a - b \in I_k$ and so $a - b \in I$. Now let $r \in R$. We have $a \in I_j$. Hence $ar \in I_j$ so that $ar \in I$. Therefore I is an ideal of R.

5. In the notation of 14.16 we have $x^2 = 1 + x$, i.e.

$$0 = x^2 - x - 1 = x^2 + x + 1.$$

Also

$$(x+1)^2 + (x+1) + 1 = x^2 + 3x + 3 = x^2 + x + 1 = 0.$$

7. Let $f(X)$ be a monic quadratic polynomial which is reducible over $\mathbb{Z}/p\mathbb{Z}$. Thus $f(X)$ is of one of two types: (i) $f(X) = (X - a)^2$ for some $a \in \mathbb{Z}/p\mathbb{Z}$, and there are p such polynomials $f(X)$; (ii) $f(X) = (X - a)(X - b)$ where a and b are *distinct* elements of $\mathbb{Z}/p\mathbb{Z}$. There are $p^2 - p$ such pairs (a, b), but the pairs (a, b) and (b, a) give the same $f(X)$. Hence there are $(p^2 - p)/2$ polynomials of type (ii). Thus the total number of reducible monic quadratics is $p + (p^2 - p)/2 = p(p + 1)/2$. The total number of monic quadratics is p^2 because they are of the form $X^2 + uX + v$ with $u, v \in \mathbb{Z}/p\mathbb{Z}$. Therefore the number of monic irreducible quadratics is $p^2 - p(p + 1)/2 = p(p - 1)/2$. Since

$p(p-1)/2$ is positive it follows that there is an irreducible quadratic polynomial over $\mathbb{Z}/p\mathbb{Z}$. Therefore there is a field with p^2 elements (14.17).

8. The equation $X^2 + 1 = 0$ has no solution in $\mathbb{Z}/3\mathbb{Z}$. Hence a is an irreducible element in R with $\partial(a) = 2$. Therefore, by 14.17 K is a field with 3^2 elements. The 9 elements of K can be denoted by 0, 1, 2, x, $x+1$, $x+2$, $2x$, $2x+1$, $2x+2$ where $3y = 0$ for all $y \in K$ and $x^2 + 1 = 0$. Let G be the multiplicative group of non-zero elements of K. Then $|G| = 8$. We have $x+1 \neq 1$; $(x+1)^2 = x^2 + 2x + 1 = 2x \neq 1$; $(x+1)^4 = (2x)^2 = 4x^2 = x^2 = -1 \neq 1$. Thus the order of $x+1$ as an element of G is none of the numbers 1, 2, 4. Since $o(x+1) \mid 8$ we have $o(x+1) = 8$. Therefore $x+1$ generates G. The elements $x+2$, $2x+1$, $2x+2$ are also generators.

Solutions 15

1. We have $q - 1 = (p^d)^r - 1 = t^r - 1 = (t-1)k$ where

$$k = t^{r-1} + t^{r-2} + \cdots + t^2 + t + 1.$$

Hence

$$X^{q-1} - 1 = (X^{t-1})^k - 1 = (X^{t-1} - 1)f(X),$$

where $f(X) = (X^{t-1})^{k-1} + \cdots + X^{t-1} + 1$.

2. We know that x is in the subfield L. Also $L \supseteq \mathbb{Z}/p\mathbb{Z}$ (9.13). Therefore L contains all polynomial expressions in x with coefficients in $\mathbb{Z}/p\mathbb{Z}$. However, every element of K is expressible in this form. Therefore $L = K$.

References

Biggs, N. L. (1985). *Discrete mathematics*. Clarendon Press, Oxford.

Burn, R. P. (1982). *A pathway into number theory*. Cambridge University Press.

Burton, D. M. (1980). *Elementary number theory*. Allyn and Bacon, Boston.

Coxeter, H. S. M. (1961). *Introduction to geometry*. Wiley, New York.

Dudley, U. (1978). *Elementary number theory*. Freeman, San Francisco.

Edwards, H. M. (1977). *Fermat's last theorem: a genetic introduction to number theory*. Springer, New York.

Hardy, G. H. and Wright, E. M. (1979). *Introduction to the theory of numbers*, 5th edn. Clarendon Press, Oxford.

Niven, I. and Powell, B. (1976). Primes in certain arithmetic progressions. *American Mathematical Monthly*, **83**, 467–9.

Rouse Ball, W. W. and Coxeter, H. S. M. (1974). *Mathematical recreations and essays*, 12th edn. University of Toronto Press.

Small, C. (1977). Waring's problem. *Mathematics Magazine*, **50**, 12–16.

Stewart, I. (1989). *Galois theory*. Chapman and Hall, London.

Index